Nursery Practices in Oil Palm

A Manual

Techniques in Plantation Science Series

Series editors:

Brian P. Forster, Lead Scientist, Verdant Bioscience, Indonesia
Peter D.S. Caligari, Science Strategy Executive Director, Verdant Bioscience, Indonesia

About the series:

A series of manuals covering techniques in plantation science that form the essential underlying needs to carry out plantation science.

The series reflects the expertise in Verdant Bioscience that underlies the plantation science activities carried out at the Verdant Plantation Science Centre at Timbang Deli, Deli Serdang, North Sumatra, Indonesia.

Titles available:

1. **Crossing in Oil Palm: A Manual** – Umi Setiawati, Baihaqi Sitepu, Fazrin Nur, Brian P. Forster and Sylvester Dery
2. **Seed Production in Oil Palm: A Manual** – Eddy S. Kelanaputra, Stephen P.C. Nelson, Umi Setiawati, Baihaqi Sitepu, Fazrin Nur, Brian P. Forster and Abdul R. Purba
3. **Nursery Screening for *Ganoderma* Response in Oil Palm Seedlings: A Manual** – Miranti Rahmaningsih, Ike Virdiana, Syamsul Bahri, Yassier Anwar, Brian P. Forster and Frédéric Breton
4. **Mutation Breeding in Oil Palm: A Manual** – Fazrin Nur, Brian P. Forster, Samuel A. Osei, Samuel Amiteye, Jennifer Ciomas, Soeranto Hoeman and Ljupcho Jankuloski
5. **Bunch and Oil Analysis of Oil Palm: A Manual** – Pujo Widodo, Fazrin Nur, Evi Nafisah, Brian P. Forster and Hasrul Abdi Hasibuan
6. **Nursery Practices in Oil Palm: A Manual** – Nur Dian Laksono, Umi Setiawati, Fazrin Nur, Miranti Rahmaningsih, Yassier Anwar, Heru Rusfiandi, Eben Haeser Sembiring, Brian P. Forster, Avasarala Sreenivasa Subbarao and Hafni Zahara

Nursery Practices in Oil Palm

A Manual

Nur Dian Laksono
Verdant Bioscience, Indonesia

Umi Setiawati
Verdant Bioscience, Indonesia

Fazrin Nur
Verdant Bioscience, Indonesia

Miranti Rahmaningsih
Verdant Bioscience, Indonesia

Yassier Anwar
Verdant Bioscience, Indonesia

Heru Rusfiandi
Verdant Bioscience, Indonesia

Eben Haeser Sembiring
PT Prime Supplier International, Indonesia

Brian P. Forster
Verdant Bioscience, Indonesia

Avasarala Sreenivasa Subbarao
Netafim, India

Hafni Zahara
Indonesian Governmental Quarantine Agency, Indonesia

CABI is a trading name of CAB International

CABI	CABI
Nosworthy Way	745 Atlantic Avenue
Wallingford	8th Floor
Oxfordshire OX10 8DE	Boston, MA 02111
UK	USA

Tel: +44 (0)1491 832111
Fax: +44 (0)1491 833508
E-mail: info@cabi.org
Website: www.cabi.org

Tel: +1 (617)682-9015
E-mail: cabi-nao@cabi.org

A catalogue record for this book is available from the British Library, London, UK.

Library of Congress Cataloging-in-Publication Data

Names: Laksono, Nur Dian, author.
Title: Nursery practices in oil palm : a manual / Nur Dian Laksono, Umi Setiawati, Fazrin Nur, Miranti Rahmaningsih, Yassier Anwar, Heru Rusfiandi, Eben Haeser Sembiring, Brian P. Forster, Avasarala Sreenivasa Subbarao, Hafni Zahara.
Description: Wallingford, Oxfordshire, UK ; Boston, MA : CABI, [2019] | Series: Techniques in plantation science | Includes bibliographical references and index.
Identifiers: LCCN 2019013221 (print) | LCCN 2019017852 (ebook) | ISBN 9781789242157 (ePDF) | ISBN 9781789242164 (ePub) | ISBN 9781789242140 (pbk : alk. paper)
Subjects: LCSH: Oil palm--Propagation. | Nurseries (Horticulture)
Classification: LCC SB299.P3 (ebook) | LCC SB299.P3 L35 2019 (print) | DDC 633.8/51--dc23
LC record available at https://protect-eu.mimecast.com/s/YQhyCZznwfLZ8iz 39zh?domain=lccn.loc.gov

ISBN-13: 978 1 78924 214 0 (pbk)
         978 1 78924 215 7 (ePDF)
         978 1 78924 216 4 (ePub)

Commissioning Editor: Rebecca Stubbs
Editorial Assistant: Emma McCann
Production Editor: James Bishop

Typeset by SPi, Pondicherry, India
Printed and bound in the UK by Severn, Gloucester

Series Foreword – Techniques in Plantation Science

Verdant Bioscience, Singapore (VBS) is a company established in October 2013 with a vision to develop high-yielding, high-quality planting material in oil palm and rubber through the application of sound practices based on scientific innovation in plant breeding. The approach is to fuse traditional breeding strategies with the latest methods in biotechnology. These techniques are integrated with expertise and the application of sustainable aspects of agronomy and crop protection, alongside information and imaging technology which not only find relevance in direct aspects of plantation practice but also in selection within the breeding programme. When high-yielding planting material is allied with efficient plantation practices, it leads to what may be termed 'intensive sustainable' production. At the same time, the quality of new products is refined to give more specialised uses alongside more commodity-based oil production, thus meeting the market demands of the modern world community, but with a minimal harmful footprint. An essential ingredient in all this is having sound and practical protocols and techniques to allow the realisation of the strategies that are envisaged.

To achieve its aims, VBS acquired an Indonesian company called PT Timbang Deli Indonesia, with an estate of over 970 hectares of land at Timbang Deli, Deli Serdang, North Sumatra, Indonesia, and the group works under the name of 'Verdant'. A central part of this estate, which will be used for important plant nurseries and field trials, is the development of the Verdant Plantation Science Centre (VPSC), to which the operational staff moved in October 2016. A seed production and marketing facility is now established at VPSC for commercial seed sales and the processing of seed from breeding programmes. The centre comprises specialised laboratories in cell biology, genomics, tissue culture, pollen, soil DNA, plant and soil nutrition, bunch and oil, agronomy and crop protection. Field facilities include extensive nurseries, seed gardens and trials (trial sites are also located at various locations across Indonesia). It is the aim of the company to use its existing and rapidly developing intellectual property (IP) to develop superior cultivars that not only have outstanding yield but are also resistant to both biotic and abiotic stresses, while at the same time meeting new market demands. Verdant not only develops and supplies superior planting materials but also supports its customers and growers with a package of services and advice in fertiliser recommendations and crop protection. This is all part of a central mission to promote green, eco-friendly agriculture.

Brian P. Forster and Peter D.S. Caligari
Lead Scientist and Science Strategy Executive Director
Verdant Bioscience

Contents

Acknowledgements

The authors are grateful to all the breeding and biotechnology teams of Verdant for sharing their knowledge and providing helpful advice in preparing this manual. The authors are extremely grateful for the inputs and advice provided by Julie Flood (CABI) especially with respect to national and international listings of risk pests and diseases subject to quarantine control.

Preface

As noted in the foreword to this series, a central objective in Verdant's mission is to develop better, more productive and more sustainable cultivars of oil palm, rubber and other plantation crops, through plant breeding. For high-yield material to reach its potential, the best growing practices need to be adopted and nursery practices are essential in this process. This is inevitably the first stage in preparing material for field planting and one that is critical but sometimes dismissed in importance because it is remote from the actual production phase. In reality it is a phase which determines the quality and performance of the adult palm. This manual describes practices conducted in the pre-nursery and the main nursery from the reception of germinated seed- or tissue-culture produced ramets to the production of high-quality field-ready plants. These practices involve nursery set-up, soil preparation, sowing, potting-on, watering, fertiliser application, pest and disease control, weeding and culling. In addition to these normal nursery practices, additional chapters are provided on quarantine nursery practices (requirements for the importation of material from sources outside the country) and DNA screening of nursery plants (pre-field planting screening for desirable traits). The manual forms part of a series in 'Techniques in Plantation Science' and may be seen as the stage describing procedures/techniques between those described in the manuals, in the same series, especially those on *Seed Production in Oil Palm* and the forthcoming *Field Trialling in Oil Palm*. Our target audiences are nursery growers, planters, students and researchers in agriculture, plant breeders and end users interested in the practicalities of producing high quality oil palm planting materials for commercial production and breeding.

Brian P. Forster and Peter D.S. Caligari

Introduction

Abstract

The performance of oil palm plantations is determined from an early stage by the quality of the planting materials. These are mainly seedlings, but ramets (usually produced from tissue culture cloning) are also used. Raising seedlings and ramets is performed in a nursery. The aim of oil palm plantations is to produce high yields, and a basic necessity is the provision of good quality planting materials. This in turn requires good nursery practices. Direct planting of oil palm seed is possible, but it cannot produce uniform materials. Nursery-grown seedlings or ramets provide healthy, strong and uniform planting materials at a suitable stage for field planting, which results in good seedling/ramet establishment and thus a high yield potential. Nursery practices are therefore a critical component of the oil palm plantation industry. In addition to commercial young plant production, seedlings and ramets are also needed for trialling (young plant screening and mature palm field performance testing) to assess progenies and breeding lines for selection and breeding and materials for pest and disease resistance or tolerance to abiotic stresses and responses to, and suitability for, changing agronomic practices and new planting materials.

1.1 History of Oil Palm Nursery Practices

The health of palms coming from the nursery has a huge effect on plant establishment once transplanted into the field. Thus, great attention needs to be paid to nursery techniques at all stages, from sowing seeds or planting ramets to the delivery of field-ready plants in the best condition possible.

Various systems have been developed in oil palm-growing countries around the world. Countries may vary in climate, soils, pest and disease incidence and management. Oil palm seedlings in Africa, for example, have specific challenges due to seasonal climate changes as compared to

Southeast-Asia, which has a more uniform climate. This, in turn, affects the occurrence of nursery diseases, and was a stimulus to initiate nursery research in West Africa (Corley and Tinker, 2015).

Early methods that were set as standards in oil palm nurseries, and which lasted until the 1960s, involved sowing germinated seeds in pre-nursery beds or pots at high density, and then transferring the young plants to a specially prepared nursery for about a year, until the seedlings reached the four- to five-frond stage. In Southeast Asia, large polythene bags were introduced in the mid-1960s to raise seedlings and this became the standard practice. Growing young oil palm plants (seedlings and ramets) in planting bags of various sizes has been tried, tested and developed. Small planting bags proved to be convenient for sowing germinated seeds and planting tissue culture-produced ramets, which are then transferred to larger bags in the main nursery. This is also cheaper than field nursery practices (described briefly below). Growing young palms in bags reduces labour costs and provides a convenient means of transporting plants from the nursery to the field. As with many plant species, this practice also reduces plant stress, as there is minimal disturbance to the plant during transfer (especially over large distances/time) and during field planting. The various stages of planting materials from seed to field-ready plants are illustrated in Table 1.1. Traditional oil palm nurseries have a high demand/use of polybags which raises concerns for the environment. Alternatives such as the use of biodegradable bags and reusable plastic trays are therefore of interest and a welcome development.

Fig. 1.1. Seed and seedling development in an oil palm nursery. a) Un-germinated seed; b) Germinated seed ready for sowing (0 weeks); c) Sowing seedlings; d) Seedlings in small bags ready for transplanting (12 weeks); e) Transplanted seedling in big bag with spacing (main nursery); f) Palms in big bag ready for field planting (36–48 weeks).

Direct planting of oil palm germinated seed in the field is possible, but there are several problems. First, there is a significant risk of damage from animals (especially insects and rodents). Secondly, the plants generated will not be uniform and abnormal palms cannot be discarded. Thirdly, there is wastage of time in crop production: palms take longer to mature and bear fruit, compared to planting field-ready plants raised in a nursery (Turner and Gillbanks, 2003; Corley and Tinker, 2015). Thus, direct field planting is generally neither practical nor economical and so is not recommended. It is instead recommended that reputable suppliers provide seed, seedlings, and/or ramets for the nursery production of field-ready plants.

Note: seeds extracted from fruits produced in commercial plantations and volunteer palm seedlings in a plantation should never be used as planting materials because commercial oil palm has a Tenera fruit form (thin shelled) which is produced by crossing a Dura (thick shelled fruit) female with a Pisifera (no shell) male. This is achieved by careful and deliberate cross-pollination by seed-producing companies and is not what occurs by random open pollination of commercial Teneras – the trait is not true breeding (see Setiawati *et al.*, 2018).

In some circumstances, bare-rooted seedlings (such as those removed from the wild in African jungles) may be transferred to the nursery. Here, sand beds, raised beds, frames and wedge practices may be used as described by Duckett (1999).

There are two commonly used oil palm nursery types: single and double stage. Single stage involves a main nursery only; double stage involves a pre-nursery and a main nursery stage. The various activities of a single- and double-stage nursery are shown in Table 1.1 and a general layout of a nursery is given in Fig. 1.2.

Table 1.1. Activities in single- and double-stage oil palm nurseries.

Activity	Single-stage nursery	Double-stage nursery
Sowing seed	√	√
Use of small bags		√
Use of large bags	√	√
Transplanting		√
Spacing	√	√
Irrigation	√	√
Pest and disease control	√	√
Culling	√	√

The decision as to which nursery system should be used depends on circumstances, which are discussed in the next section.

1.2　Importance of Nursery Best Management Practices

Oil palm is a crop that comes into maturity at about 20–25 months after field planting. The earlier it is in production the earlier the profits can be reaped. Peak yield occurs at 4–5 years after first harvest or 9–18 years after field planting (Alam *et al.*, 2015). Good nursery management provides strong and healthy plants (seedlings or ramets), which lead to good field establishment after planting and, in turn, promotes early flowering and fruit production and thus early and high yields. This then leads to a long productive life span of the plantation. Substandard planting materials will have long-term consequences for yield, i.e. throughout the lifetime of the plantation, which may be as long as 25 years (Heriansyah and Tan, 2005).

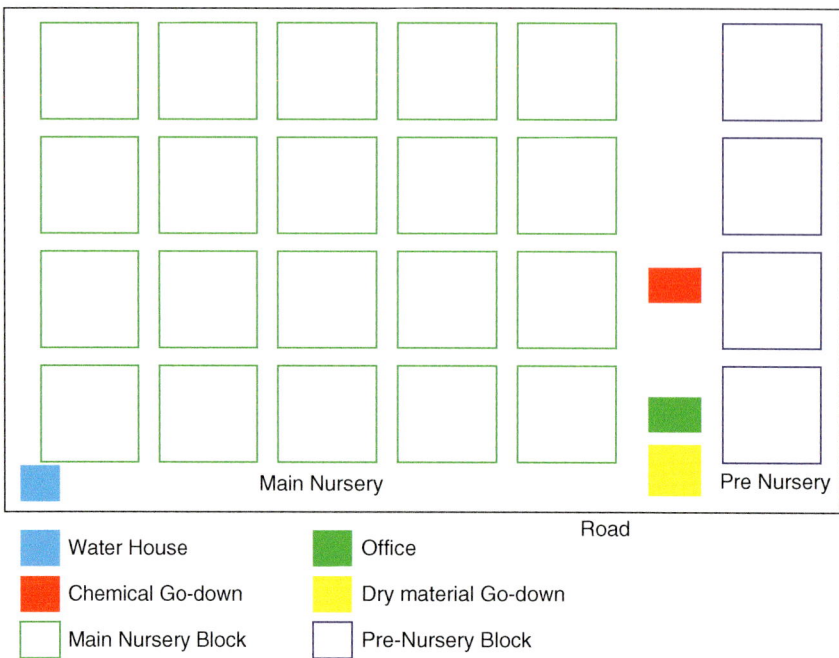

Fig. 1.2. Components of an oil palm nursery layout: blocks in the main nursery and pre-nursery, road access, dry and chemical go-down/store, water house, road construction and office. Other off-site facilities include soil supply, incinerator and offices.

The production of superior oil palm planting material is dependent on all stages and all procedures in the nursery. The procedures of an oil palm nursery need to be followed stringently.

a) **Seedlings.** Area preparation, single-/double-stage nursery, nursery maintenance (manure and fertiliser application, watering, culling and weeding), pest and disease control.

b) **Ramets.** Activities are the same as seedlings, the difference is in shading young ramets and a longer pre-nursery period for acclimatisation from tissue culture conditions.

Oil palm seedlings are developed from germinated seeds arising from controlled pollination (either for commercial production or breeding purposes). Ramets (plantlets) are clones produced from tissue culture. Both are raised using the standard operating procedures (SOPs) of nursery husbandry. Basic requirements are adequate sunshine/shading, potting soil, water, supplementary nutrients and control of pests and diseases (Duckett, 1999). The production of good quality planting materials is the main objective; the end product is field-ready planting material that is uniform, healthy and vigorous. It is possible, if needed, to hold palms in the nursery until the optimal planting time, as determined by the needs of the planter (Heriansyah and Tan, 2005).

As mentioned above, there are two types of oil palm nursery: single stage and double stage. The main advantage of the single-stage nursery is a reduction, by about two months, of the overall time spent in the nursery (Bevan and Gray, 1969; see also Table 1.1). The single-stage nursery gives larger palms for field transplanting at 13 months from germination. However, such operations require high labour and water requirements in the first four months, in turn requiring close supervision, especially at the initial plant growth phase. Such nurseries also require a larger total area (Rankine and Fairhurst, 1998).

Traditional nursery practices use large amounts of polybags. Single-use plastic is now being replaced by more environmentally friendly practices such as the use of biodegradable bags and re-usable plastic trays. Raising oil palm seedlings in a nursery pot tray system was reported by Chee *et al.* as early as 1997. This system can be used repeatedly to raise up to 20 cycles of oil palm seedlings, and the pot trays used can be built from recycled plastic. Mathews *et al.* (2010) reported that this method is being taken up by most oil palm companies. In the pre-nursery stage this method is more efficient: the area needed per seedling is less (up to 169 seedlings/m^2), less planting medium required, ease of transfer and more sustainable.

The large polybags that are currently used traditionally in the Main nursery stage can be replaced by biodegradable bags. Muriuki *et al.* (2013) and Bilck *et al.* (2014) describe benefits of biodegradable bags, a major benefit, aside from sustainability, is that the bag containing the seedling can be direct planted, thus avoiding root disruption. Another sustainable issue in the nursery is the use of topsoil. Small planting bags usually require about 1.2 kg of soil per seedlings, but this can be reduced to a tenth if pot tray systems are used (Mathews, *et al.*, 2010). However, the quantities of topsoil required in the Main nursery is considerably larger, up to 20-25 kg/planting bag (CIRAD, 2008). Alternatives are now being used. Cocopeat, compost, POME (palm oil mill effluent) are materials that may be used to mix with soil. Mohammad *et al.* (2012) reported the effective use of 100% cocopeat, and Lord and Betitis (2007) showed no differences between 100% soil and 100% compost (POME) derived from oil palm empty fruit bunches (a bi-product from oil palm mills). Thus, topsoil usage can be reduced or even eliminated.

Other important and routine aspects in nursery plant care include fertiliser application, watering, monitoring and controlling pests and diseases, as well as culling of off-type plants. It is recommended that fertiliser treatments of young oil palm plants are a combination of organic fertiliser and compost applications. Organic fertilisers are preferred over inorganic fertilisers because they are milder and there is less risk of damage caused by more aggressive inorganic chemicals. In addition, organic fertilisers improve the physical, chemical and biological properties of the soil and can provide carbon compounds that may also improve the physical properties and microbial community of the soil (Ariyanti *et al.*, 2018; see also Chapter 6).

The availability of water is essential to meet the needs of actively growing young plants. Water deficiency has negative effects on plant metabolism, physiology and morphology, as well as disease and pest susceptibility, and can be hugely damaging to young plants. If neglected, plant death can result (Sun *et al.*, 2011; see also Chapter 7).

Young plants are also vulnerable to a range of pests and diseases. These can cause severe losses if not recognised at an early stage and immediate action taken. Thus, prevention and control of pests and diseases is a major concern (Mathews *et al.*, 2010; see also Chapter 10), especially in quarantine nurseries (see Chapter 12).

New practices in the nursery include pre-field planting screening. Screening is traditionally done in the nursery for phenotypic off-types (e.g. slow growth and development, crinkled leaf, wide internode, chimera, etc., see Chapters 5 and 8) which are culled before transfer to the field. Culling is a major nursery activity and is typically done three times at 3, 6 and 9 months. Common culling rates are between 10 – 25%. Lower than 10% is considered risky and greater than 25% indicates problems in the material (either inherent or due to poor nursery practices). Other screening includes seedling screening for response to *Ganoderma*, a major disease of oil palm. Breeders and seed producers are consequently keen to develop materials that exhibit resistance to this disease. Seedling screening provides an early and quick method of assessing disease resistance/susceptibility. A manual on nursery screening of *Ganoderma* response is available in this series (Rahmaningsih *et al.*, 2018).

Genomic selection (use of DNA diagnostics) is now a reality in oil palm breeding and has application in quality control in commercial plant production. In relation to DNA testing, leaf samples can be taken from nursery plants and transferred to the laboratory where DNA is extracted and screened for specific genes/traits, or for the correct genetic background in the case of legitimacy testing and in the case of ramets for somaclonal variation. The primary example of the application of such testing is for fruit type. The shell thickness gene controls fruit type, which is the most commercially important trait in oil palm. Commercial oil palm has the Tenera (*Sh/sh*) thin-shelled phenotype and is produced from crosses between thick-shelled Dura (*Sh/Sh*) females and no-shelled Pisifera (*sh/sh*) males (Setiawati *et al.*, 2018). DNA testing can discriminate between Tenera, Dura and Pisifera genotypes and thus selections can be made in progenies, for example the removal of Dura contaminants among commercial Tenera seedlings (Sambanthamurthi *et al.*, 2009) and the removal of Pisiferas which are sterile. Thus field trialling can focus on relevant fertile Dura or Tenera germplasm (see Sitepu *et al.*, 2019 this series; see also Chapter 11). Fruit colour change, black-red (*Nig*) or green-yellow (*Vir*) is another trait that can be easily determined by DNA analysis. As with shell type, this trait can only be seen at maturity, thus DNA screening in the nursery not only picks out desired types but saves five years in determination.

Other considerations for the nurseryman are checks that seeds are from the right source (legitimate), that there is ample provision of financial capital for buying sprouted seeds or tissue culture-produced ramets, proper handling of germinated seeds/ramets and the application of improved agronomic practices in the management of the nursery (Yusuf *et al.*, 2014).

The nursery practices outlined above are based on local or within-country supply of seeds and ramets. Oil palm breeders continuously access materials from other countries, and their entry into the country is regulated by strict import and quarantine controls. For seed, seedling and ramet materials, this requires growing in a specialised quarantine nursery where materials can be sampled and inspected regularly over a prescribed quarantine period by official quarantine officers, before release. These issues are described in detail in Chapter 12.

References

Alam, A.S.A.F, Er, A.C. and Begum, H. (2015) Malaysian oil palm industry: Prospect and problem. *Journal of Food, Agriculture and Environment* 13(2),143–148.

Ariyanti, M., Santi, R., Intan, R.D. and Aldo, F. (2018) The growth response of oil palm seedling at main nursery against watering at different volume and frequency and against provision of compost. *International Journal of Sciences: Basic and Applied Research* 37(3), 226–233.

Bevan, J.W.L. and Gray, B.S. (1969) *The Organization and Control of Field Practice for Large-Scale Oil Palm Plantings in Malaysia.* The Incorporated Society of Planters, Malaysia.

Bilck, A.P., Olivato, J.B., Yamashita, F., and de Souza, J.R.P. (2014) Biodegradable bags for the production of plant seedlings. *Polimeros.* 24(5): 547–553.

Chee, K.H., Chiu, S.B. and Chan, S.M. (1997) Pre-nursery seedlings grown on pot trays. *The Planter* 73(855): 295–299.

CIRAD (2008) *Germinated Oil Palm Seeds: Recommendations for Prenursery and Nursery Management.* Available at: https://www.palmelit.com/en/content/download/4239/32933/version/8/file/Booklet-of-recommandations-for-prenursery-and-nursery-management-oil-palm-seeds-CIRAD.pdf (accessed 30 October 2018).

Corley, R.H.V. and Tinker, P.B. (2015) *The Oil Palm*, 5th edn. Wiley Blackwell, UK, p. 639, ISBN: 978-1-405-18939-2.

Duckett, J.E. (1999) *A Guide to Oil Palm Nurseries.* The Incorporated Society of Planters, Malaysia.

Heriansyah, B. and Tan, C.C. (2005) Nursery practices for production of superior oil palm planting materials. *The Planter* 81(948), 159–171.

Lord, S. and Betitis, S. (2007).The role of EFB compost and its effect in New Britain Palm Oil Ltd. (NBPOL) nurseries. In Proceedings of the PIPOC 2007 International Palm Oil Congress (Agriculture, Biotechnology & Sustainability), 26–30 August 2007, Kuala Lumpur Convention Centre, Malaysia, pp. 599–624.

Mathews, J., Tan, T.H., Yong, K.K., Chong, K.M., Ng, S.K. and Ip, W.M. (2010) Managing oil palm nursery: IOI's experience. *The Planters* 86(1016), 771–785.

Mohammad, M.K., Kamarozaman, A.A., Arifin, I. and Nasir, A.R.M. (2012) Evaluation of several planting media for oil palm (*Elaies guineensis*) seedlings in main nursery. In Kadir W.R. *et al.* (Eds) Soil Science Conference of Malaysia: Soil quality towards sustainable agriculture production, 10–12 April 2012, Kota Bahru, Malaysia, p. 542.

Muriuki, J.K., Kuria, A.W., Muthuri, C.W., Mukuralinda, A., Simons, A.J. and Jamnadass, R.H. (2013) Testing Biodegradable Seedling Containers as an Alternative for Polythene Tubes in Tropical Small-Scale Tree Nurseries. *Small-scale Forestry* 13(2): 127–142.

Rahmaningsih, M., Virdiana, I., Bahri, S., Anwar, Y., Forster, B.P. and Breton F. (2018) *Nursery Screening for* Ganoderma *Response in Oil Palm Seedlings: A Manual. Techniques in Plantation Science*. Forster, B.P. and Caligari, P.D.S. (eds). CAB International, Wallingford, UK, p. 96.

Rankine, I. and Fairhurst, T. (1998) *Oil Palm Nursery – Field Handbooks*. Potash and Phospate Institute, Singapore.

Sambanthamurthi, R., Rajinder, S., Ahmad, P.G.K., Meilina, O.A. and Ahmad, K. (2009) *Opportunities for the Oil Palm via Breeding and Biotechnology in Breeding Plantation Tree Crops: Tropical Species*. Springer-Verlag, New York.

Setiawati, U., Sitepu, B., Nur, F., Forster, B.P. and Dery, S. (2018) *Crossing in Oil Palm: A Manual. Techniques in Plantation Science*. Forster, B.P. and Caligari, P.D.S. (eds). CAB International, Wallingford, UK, p. 96.

Sitepu, B., Setiawati, U., Nur, F., Laksono, N.D., Anwar, Y. *et al.* (2019) *Field Trials in Oil Palm Breeding: A Manual. Techniques in Plantation Science*. Forster, B.P. and Caligari, P.D.S. (eds). CAB International, Wallingford, UK, in press.

Sun, C.X., Hong-xing, C., Hong-bo, S., Xin-tao, L. and Yong, X. (2011) Growth and physiological responses to water and nutrient stress in oil palm. *African Journal of Biotechnology* 10(51), 10465-10471.

Turner, P.D. and Gillbanks, R.A. (2003) *Oil Palm Cultivation and Management*, 2nd edn. The Incorporated Society of Planters, Malaysia.

Yusuf, S.A., Muhammad, I.A.Q. and Justina, O.L. (2014) Determinants of risk and uncertainty in oil palm nursery. *Journal of Economics and Sustainable Development* 5(11), 174–186.

Health and Safety Considerations   **2**

Abstract

Standard operating procedures and safety protocols are needed in all activities in the nursery. The official standards can vary depending on the country and local regulations, but high standards should be maintained whatever requirements are officially needed. Identification and elimination of hazards and risks, followed by developing specific safety procedures, and procedures for preventing and responding to workplace accidents and injuries are important features in establishing an effective occupational health and safety programme. Guidelines in health and safety issues relating to nursery practices in oil palm are given below.

2.1 The Importance of Health and Safety

Although the work risk in the nursery is not as high as in the field, the health and safety of workers is paramount and, in addition, attention should always be paid to environmental considerations. A key consideration is working outside and being subject to prevailing weather, sun and pests. Although most work is in the nursery itself, various activities outside the plant area are needed, such as the preparation of sprays to control pests and diseases, maintenance of watering systems and incineration of waste.

Personal protective equipment (PPE) is essential in protecting workers from injury and other health and safety threats. Typical PPE for nursery activities are listed below.

- **Hats.** Wide-brimmed hats should be worn as a protection from the sun, even on cloudy days.
- **Sun block** should be supplied and advice given on its application.
- **Safety boots.** These should be worn in every activity in the nursery due to various hazards, e.g. walking and working on wet surfaces, protection

from thorns, bites and stings from ground animals such as snakes and scorpions, and broadcasted chemical materials such as insecticide which is scattered on the surface of the ground.

- **Gloves.** These need to be worn to avoid hand injuries for some activities such as planting (oil palm plants have spines) and spraying (avoid skin contact with chemicals).
- **Safety goggles or similar eye protection** are advised to avoid eye injury caused by splinters and splashes. Eye protection must be worn when working with chemicals such as pesticides, insecticides and fungicides, and when spraying to control pests and diseases.
- **Safety respirator mask.** This should be worn when activities involve chemical materials or any hazards that can be inhaled, such as pesticides when spraying and pollen when crossing.
- **Aprons.** Should be worn when spraying with chemicals.

Health and safety of workers and the protection of the environment are paramount. Work risks in the nursery can be high if standard health and safety issues are not observed. Hazards vary depending on the activity.

Occupational health and safety in the nursery are classified and described for various operations and updated by the International Labour Organisation (ILO).

- Physical hazards:
  - operational and workplace hazards;
  - machinery and vehicles;
  - confined and restricted space entry;
  - risk of fire and explosion.
- Biological hazards.
- Chemical hazards.

Health and safety training and refresher training are essential for worker safety. The ILO and the Roundtable on Sustainable Palm Oil (RSPO) are international organisations that have concerns for worker safety and raise additional considerations about health and safety (see e.g. ILO, 2001; RSPO, 2013). ILO and RSPO issues include:

- that equipment works properly and is checked before use;
- safe chemical use (especially pesticides);
- SOPs;
- working alone;
- emergency procedures, first-aid box;
- awareness of nuisance insects (e.g. mosquitoes) and other animals (e.g. snakes);
- training and refresher training.

References

ILO (2001) *Convention 184: Convention Concerning Safety and Health in Agriculture.* Available at: https://www.ilo.org/public/english/standards/relm/ilc/ilc89/pdf/c184. pdf (accessed 25 July 2018).

RSPO (2013) *Principles and Criteria for the Production of Sustainable Palm Oil.* Available at: https://rspo.org/publications/download/4b4296c7bb85cb3 (acessed 25 July 2018).

Nursery Set-up

3

Abstract

All work activities in the nursery need preparation and this begins with setting up the nursery. In this chapter, we discuss how to select a good site for the nursery with respect to critical features (topography, soil supply, water supply, drainage, protection), which nursery type is best, land preparation, construction and preparations to begin nursery activities. In addition, 'off-site' facilities needed to support nursery work are listed, such as an incinerator, soil source and go-down/storage.

3.1   Site Selection

The following should be considered:

a) **Location.** The nursery is normally permanent, but may also be temporary. In both cases it should be close to the area to be planted.

b) **Capacity.** The number of palms that need to be grown (this affects the area needed).

c) **Irrigation.** Can an irrigation system be laid out and is there a nearby source of good quality water? Be especially aware of water availability during dry seasons. For example, site the nursery near a stream, creek or a river that runs throughout the year. Permission may be required to use local water sources.

d) **Flooding.** Make sure that the area is not prone to flooding, either from overflow from streams and rivers or after heavy rain.

e) **Topography.** The location should be flat and open (full sun).

f) **Soil.** There should be a local and plentiful supply of good topsoil (loamy soil) where planting bags can be labelled or alternatively the topsoil should be near enough to the nursery for easy transport.

g) **Road access.** There should be good road access to the nursery site.

h) **Animals.** The area should be free from wild animals and livestock. Be aware of and respect local land area use and designations such as conservation areas and protected species.

i) **Security.** The location needs to be secure against theft and vandalism and nuisance animals.

See also Duckett (1999), CIRAD (2008), Tolan Tiga Indonesia (2011) and Meadows and Litz (2016) for further information on setting up a nursery.

3.1.1 Water supply

Water requirements, both quality and quantity, are a priority in determining site selection and preparation. Young oil palm plants need 0.5 l of water each day per small planting bag, and 2 l per large planting bag (Muhamad *et al.*, 2014). Natural ponds or lakes, streams, rivers or underground reserves can be used as water sources. Intake pipes and pumps are normally placed near to the water source. A backup system should be considered, particularly in isolated areas or areas of low or unreliable rainfall. Other artificial water sources may be developed such as earth dams and water reservoirs, as described by Duckett (1999).

3.1.2 Topography (terrain)

The selected area for the nursery should be relatively flat with slopes between 0–30º. A good option is to have the lower area of the nursery close to the source of water. In Indonesia and other countries in Southeast Asia there are distinct dry periods from two to six months in a year, thus the water supply needs to be reliable in these dry seasons. A recent innovation is the use of drones to carry out aerial surveys of potential land areas using spatial imaging methods or geographic information system imaging and software (Nordiana *et al.*, 2013).

3.1.3 Area

The selected area may be surrounded by other crops or mature oil palm. These should not cause substantial shading of the nursery as this will cause etiolation (yellowing and elongation) of nursery plants. It is recommended that there is an open space of at least 20 m wide around the boundary (Fig. 3.1) to the nursery.

The required size of the nursery depends on the number and density of planting bags per ha. A good large bag density is 13,000 planting bags per ha with 0.91 m triangular spacing. This density is desirable because it:

- minimises etiolation of young plants (seedlings or ramets);
- provides good access for fertiliser application;
- provides a safety buffer in the event of there being delays in pre-paring the field for planting, as nursery plants may be maintained for longer; and
- provides good growing conditions to 'supply' field-ready plants (replace-ments kept in the nursery to be supplied in the event of plant loss after planting, e.g. due to pests, disease, damage).

Fig. 3.1. Aerial view of a nursery a) Water supply (pond); b) Pump house; c) Go-down storage buildings; d) Pre-nursery block; e) Main nursery block. Note the boundary should be wide enough to prevent shading of nursery plants.

3.1.4   Accessibility and nursery roads

Roads within the nursery need to be planned carefully, bearing in mind the type of irrigation and block layout of the nursery (this is normally a grid system). Roads within the nursery should be no less than 3 m wide to allow access by tractor or any other carrier vehicle. A good access road to the nursery should be sufficiently wide to allow vehicles to pass. There is usually a peak planting period (normally coinciding with the rainy season) and therefore the nursery needs to be prepared for intensive activity asso-ciated with the movements of vehicles and materials as well as the work that needs to be supervised (See Fig. 3.1 with respect to road layout within the nursery).

3.1.5   Drainage

Flooding will damage young palms in the nursery (it also causes problems for buildings for storage or go-downs), which leads to palm stress and nutrient imbalance. Sites with poor drainage should be avoided, and this may be assessed by consideration of topography and waterways and water supply points as described above. Poor drainage can be alleviated by the use of suitable drainage canals (Fig. 3.2).

Fig. 3.2. Nursery drainage canal (typically 30 cm for width and 50 cm for depth).

3.2   Nursery Preparation

The preparation of the area for a nursery is important to allow optimum palm growth, maintenance of nursery site, unimpeded access and hygienic growing conditions.

3.2.1   Nursery tools and materials

The tools for nursery preparation (Fig. 3.3 shows some examples) should be prepared to enable set-up and to manage the nursery properly. Basic tools and materials include: hoes, shovels, diggers, wheelbarrows, watering cans, spraying tools, small planting bags (18 cm × 21 cm × 0.1 cm), large planting bags (38 cm × 55 cm × 0.15 cm), fertilisers and chemicals (e.g. pesticides and fungicides).

Environmental issues are leading to more sustainable and eco-friendly practices, this includes the move away from single-use plastic polybags. Pot tray systems are now commonly used in nursery plantations (Mathews *et al.* 2010). These trays can be used several times at the pre-nursery stage and replace small (single-use) polybags. Chee *et al.* (1997) were first to report the pot tray system which has several advantages: the tray can be used repeat-edly, require less soil, easy to handle, easy to maintain and can be re-cycled.

Fig. 3.3. Equipment storage.

Fig. 3.4. Pot trays system at the pre-nursery stage.

In addition the trays can be placed on benches for easy access (Fig. 3.4). Pot trays used in the pre-nursery have a smaller size than traditional planting bags, hence the number of seedlings per area is higher (up to 169 seedlings/m2). The size can vary depending on requirements and planting medium, i.e. 69 mm × 60 mm × 120 mm or 58 mm × 58 mm × 120 mm.

Large polybags in main nursery stage can now be replaced with bio-degradable bags. Muriuki *et al.* (2013) and Bilck *et al.* (2014) have demonstrated the use of biodegradable bags in the production of seedlings in other plant species. Biodegradable bags also have the advantage that they can be transplanted directly into the soil without removing the bags thus reducing root damage and there is no plastic waste.

Yusuf *et al.* (2014) suggest lockable stores or cabinets for tools, equipment and for chemicals and fertilisers. Mutert *et al.* (1999) suggest that herbicides must be clearly marked and kept separately from insecticides, fungicides and foliar fertilisers to prevent contamination and incorrect handling. An incinerator for waste disposal is recommended and should be located near the nursery to prevent the spread of contamination from infected and culled nursery plants (see Section 3.5 below).

3.2.2   Clearing the site

After the nursery site has been selected, the next step is to clear the area. All trees need to be felled and all other plants and plant debris removed including tree stumps, logs, branches, leaves, etc. Clearing should be completed at least two months before the first germinated seeds or ramets are expected to arrive. (**Note:** for Indonesia and many other countries there is a zero-burning policy in clearing areas for oil palm nurseries and plantations – Noor, 2003).

3.2.3   Fencing the site

The nursery must be secure from theft and ingress of local animals, such as cattle or any other local threats. Thus the nursery must be fenced with a lockable access gate. Two types of fences are commonly used.

- **Conventional fence.** The specifications for a conventional fence depend on the species of animal that need to be kept out. For example, a four-strand wire fence, with wires spaced at 0.3, 0.6, 0.9 and 1.2 m from the ground should be adequate to control cattle, sheep and goats. An alternative is a wire-net fence (see Fig. 3.5). The fence needs to be strong enough so that animals cannot push it over. **Note:** in some areas barbed wire fencing may not be allowed.
- **Electric fence.** Where there is a range of wild mammalian pests, an electric fence is potentially a good option. Wire of 250–300 kg breaking

Fig. 3.5. Wire-net fence.

strain has been found to be the most suitable (Duckett, 1999). The height of the electrified wires is critical, e.g. 10 cm repels porcupines, but more than one strand may need to be electrified.

3.2.4   Irrigation preparation

The irrigation systems most commonly used for nursery palm plants (seed-lings and ramets) are overhead sprinklers at about 2 m in height, or flat sprinkler tubes with two rows of holes in the upper surface to spray a fine mist at opposite angles when the tube is pressurised. Hand watering systems may be adequate in small (less than 1 ha) nurseries.

The efficiency of sprinklers is poor. There is a lot of water wastage, and improper timing and inadequate water supply leads to plant stress. One technology that can improve watering efficiency is drip irrigation (Fig. 3.6). Drip irrigation of oil palm nursery plants involves precise delivery of suffi-cient smaller quantities of water and fertilisers, uniformly in each planting bag substrate, through emitters placed along the water delivery lateral line, with a higher frequency of application than other systems such as fixed sprinkler systems, sprinkler hoses, rain guns, etc. The benefits of drip irrigation are: a) efficient use of water; b) improved fertiliser application by fertigation; c) reduced operational labour; d) decreased energy requirements; e) avoidance of soil prob-lems; f) enhanced plant growth; and g) faster and higher return on investment.

Netafim has drip irrigation that is suitable for an oil palm nursery, it is called PalmDrip™ and is described below.

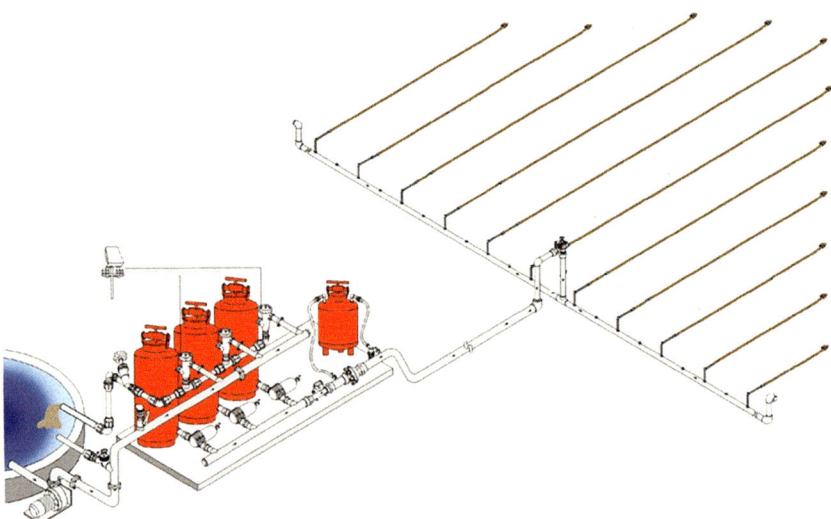

Fig. 3.6. Example of a drip irrigation (PalmDrip™) layout for oil palm pre- and main nursery.

Components of drip irrigation

Pipeline (main line, sub-main line and emitters), valves (manual and automatic) to regulate the water flow, filtration unit. A fertigation dosing unit (NetaJet™) at the head controls the amount of water and fertiliser used, and an anti-siphon unit prevents backflow of fertilisers or other chemicals into the water source (Fig. 3.6).

System layout

Palm plants are arranged in blocks in the nursery. PalmDrip™ irrigation systems are normally divided into blocks to reduce the lengths of lateral pipes required and to promote a uniform emitter discharge rate throughout the nursery area. Within each block there is a manifold that supplies water to the drip laterals, which in turn feeds the drippers mounted on the laterals. The manifold is connected to the sub-main, which in turn is connected to a main line. The nursery is split into blocks, with a manifold installed at the head of each block. Several sets of blocks of 1 ha each are often installed along the length of the nursery. Fig. 3.7 shows a typical block layout and drip irrigation system for an oil palm nursery.

Drip irrigation systems for oil palm nurseries use thick-walled hoses, spaced 1.56 m apart, commonly referred to as drip laterals. Each drip lateral feeds two rows of planting bags (Fig. 3.8). The commonly used drip lateral

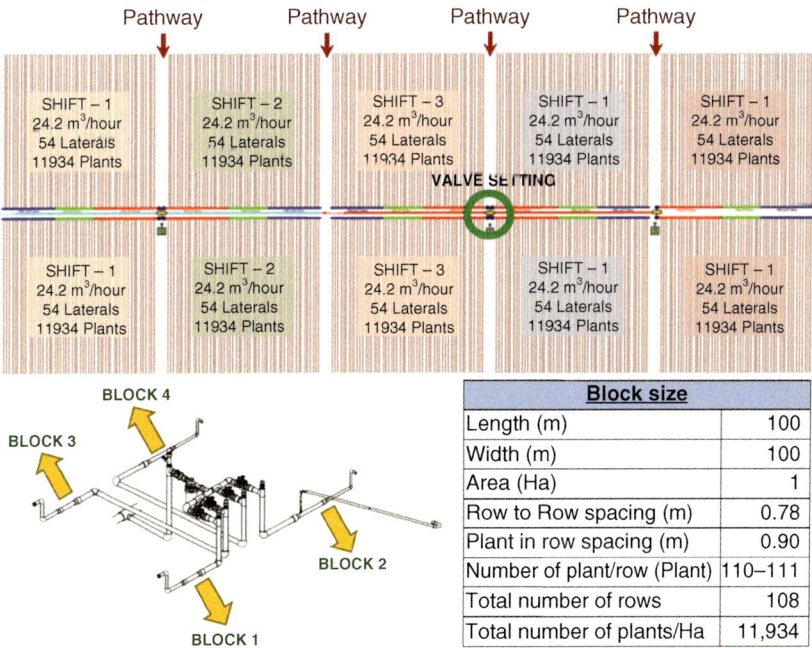

Fig. 3.7. Drip irrigation system design with a manifold and sub-main located at the head of each block.

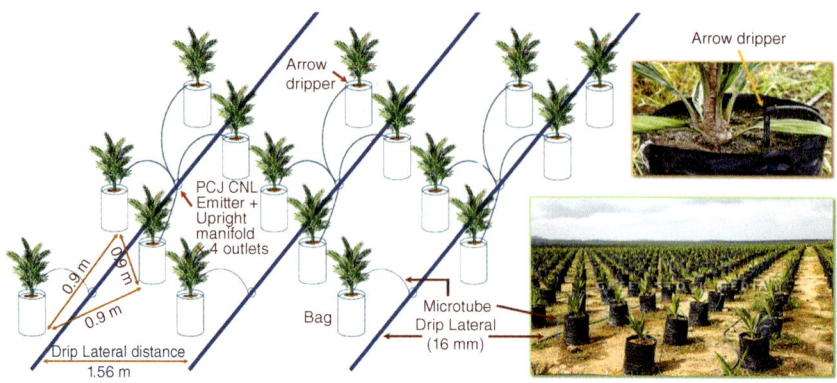

Fig. 3.8. Physical configuration of palm rows, drip laterals and emitters.

has a 16 mm outer diameter and 13.6 mm inside diameter. The thickness of the drip lateral is 1.2 mm. In some designs, a 20 mm outer diameter (17.6 mm inner diameter) is required. A thicker lateral is generally used for larger lateral diameters. Properly cared for drip laterals allow long life, as long as 10 years or more.

Emitter type
Two types of emitters are used, namely, PCJ-CNL online emitters and arrow drippers (Fig. 3.9). The emitter delivers precise and equal amounts of water over a broad pressure range of 0.7 to 4.0 bars. The filtration area is 2 mm². Uniformity of water and nutrient distribution along the laterals is 100%.

The manifold emits 8 l/hr, discharged and split into four drip outlets. Thus, each PCJ CNL online emitter feeds four arrow drippers and each arrow dripper is inserted in one planting bag delivering 2 l/hr (Fig. 3.9).

The drippers are highly UV resistant as well as resistant to standard nutrients used in agriculture. The PCJ online emitter meets ISO 9261 standards. Drippers can be positioned exactly where required. The number of drippers can be increased so as to increase water quantity supply, aimed at meeting optimum plant growth rates. The arrow dripper possesses 'turbulent labyrinth' wide water passages for highest reliability and efficiency. Arrow drippers serve as an accessory to PCJ CNL drippers for better and more accurate installation. The dripper continuously self-flushes debris throughout operations, not just at the beginning or end of a cycle, ensuring uninterrupted dripper operation.

System capacity
The PalmDrip™ system should have a design capacity adequate to satisfy the peak irrigation water requirement of each plant in a planting bag and

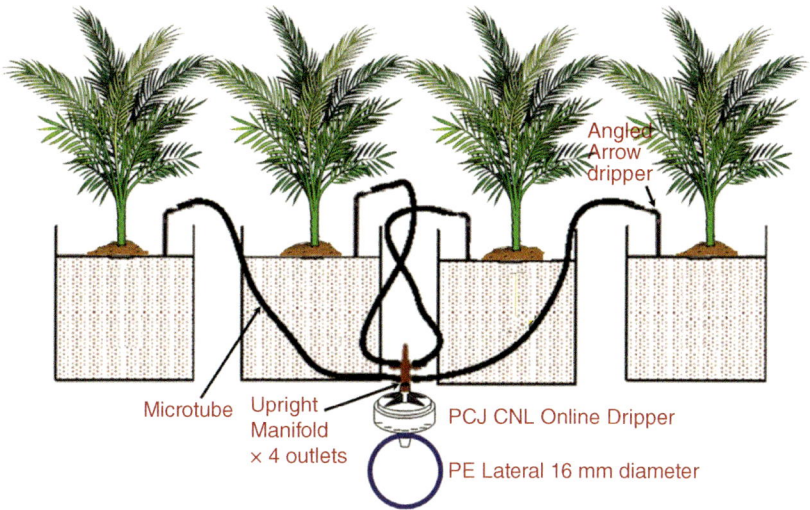

Fig. 3.9. Emitter type and layout.

all plants to be irrigated within the nursery block. The capacity should include an allowance for water losses that may occur, if any, during application periods. The system should have the capacity to apply a stated amount of water to the nursery block in a specified net operation period. The system should have a minimum design capacity sufficient to deliver the peak daily irrigation water requirements in about 90% of the time available or not more than 22 hours of operation per day. Fig.3.7 provides a calculation of water requirement with regard to the system layout for a 1 ha block.

Design flow rate
The PalmDrip™ system should be set up to provide sufficient water during the period of maximum plant water use. The design flow rate depends on the area irrigated, the irrigation set time and the design depth of application. Flow rate is related to watering activity, which is explained in more detail in Chapter 7.

3.3   Single- or Double-stage Nursery

The decision to have a single-stage or double-stage nursery will be a matter of personal choice depending on the specific situation encountered (e.g. a double-stage nursery is recommended for large-scale plantings of >500 ha).

3.3.1 Single-stage nursery

This system utilises only large planting bags, and germinated seeds or ramets are planted directly into these bags in the same manner as planting out in the pre-nursery planting bags. Some advantages of this system are as follows:

- **Fast palm establishment.** There is no movement until field planting, thus the root development is not disturbed and can establish faster. There is no/little transplanting shock when potting-on from small to large planting bags.
- **Less expensive** than double-stage because the number of activities (labour usage) is less than in the double-stage nursery (e.g. no potting-on, holing, etc.).

Disadvantages include the following;

- **Lack of flexibility.** It is necessary to have the full nursery infrastructure ready, large planting bags filled and irrigation of the full nursery area functional from seed/ramet delivery.
- **Wastage of water.** For the initial period of two to three months (the equivalent of the pre-nursery period) the irrigation system has to operate over the full nursery area. This requires a greater volume of water and additional engine fuel as well as wear and tear on the pumping and plumbing equipment.
- **Greater labour inputs.** These are needed for activities such as fertiliser application and pest and disease control, as required.
- **Space constraints.** If consecutive season plantings are carried out and there is a delay in any single year's planting programme and young field-ready plants cannot be cleared from the nursery in time, there is no space for new season seed/ramet deliveries.
- **Slow culling.** Rounds of culling involve covering large areas and therefore take far longer than in double-stage nurseries.
- **Single-stage nurseries are not recommended** for large-scale planting which requires large areas to be prepared, high costs and little flexibility of nursery space usage.

3.3.2 Double-stage nursery

This system involves two planting stages, pre-nursery (or rametry) and main nursery. The pre-nursery stage involves sowing germinated seed and growing these on for two to three months. The palms are then potted-on into large planting bags in the main nursery where they remain for a further 7–10 months before field planting. In some cases, ramets (tissue culture-produced clones) are used instead of germinated seeds (see Section 3.4.2). The flow of work activities is shown in Fig. 3.10.

Fig. 3.10. Flow of work in a double-stage nursery, starting from planting seeds in small bags in the pre-nursery, then transplanting to the main nursery, and then spacing the large bags and maintenance of the plants until they are ready for field planting.

Double-stage nurseries are generally preferred, as they require less space and irrigation, and allow for more efficient upkeep, observation and selection (culling of off-types). However, the double-stage nursery involves the potting-on of pre-nursery palms and ramets to the main nursery, which may cause transplanting shock if not done properly.

Advantages of double-stage over single-stage nurseries include the following:

- **Low water requirements.** Less water is needed for the first two to three months, thus providing savings in maintenance costs. Watering is commonly done by hand, but drip irrigation provides greater efficiency.
- **Observations at the critical early stage of plant establishment are easier and more effective.**
- **Easy to treat young plants for pests and diseases and to water them due the smaller area and ease in labour movement/activities.**
- **Culling can be carried out quickly and easily prior to potting-on into large bags.** Thus providing greater efficiency in the use of nursery space.
- **Staggered workload.** The preparation (soil filling) of large planting bags in the main nursery can be staggered over time, thus reducing pressure on workloads and required labour.
- **Greater flexibility.** If transfer to the field is delayed, the palms for the next year's planting can be held in the pre-nursery for several months until the main nursery is cleared.

Disadvantages include the following:

- **Additional work.** Additional work is required in setting up the pre-nursery and potting-on into large planting bags.
- **Risk of potting-on shock.** Transplanting from small to large planting bags has a risk of transplanting shock, which will set back or kill the palm. Potting-on needs to be done carefully.

3.3.3   General recommendations

The decision on nursery type to be adopted is based on needs, such as the periods of time in each nursery stage, number of palms/ramets, type of irrigation, flexibility, labour, etc.

If a nursery is required for several years and needs to handle a large number of plants, it is recommended to adopt a double-stage system. Double-stage nurseries provide numerous advantages such as safeguards and increased flexibility, particularly when drip irrigation is utilised.

Single-stage nurseries are only recommended when a relatively small nursery (under 2.5 ha) is needed for a short length of time (one year).

3.4   Trial Material

3.4.1   Labelling

Trial activities, especially for breeding, involve various progenies (see Sitepu *et al.*, 2019, this series). Transfer and transplanting in the field requires care and supervision because it is easy to make mistakes, such as mixing up palms between progenies to be tested. It is essential to label or field code the palms/planting bags.

When labelling palms (or ramets), the information should include a desciption of the progeny, consigment, origin, etc. Labels must be waterproof and fixed (not easily detached, see Fig. 3.11), and placed on the planting bags prior to sowing. Labelling during potting-on and transplanting is described in Chapters 4 and 5.

3.4.2   Preparation for ramets

Ramets have no kernel food reserve, thus they grow slower than seedling palms. They do not form normal (palm) fronds, and are prone to desiccation when exposed to dry air (Corley and Tinker, 2015). Thus, maintaining

Fig. 3.11. Labels on small and large bags.

a condition of high humidity is essential in the hardening of ramets. Ramets are produced from tissue culture where they grow in high humidity, sterile and protected conditions (normally in a glass or plastic vessel). In some tissue culture methods, root production is poor and the roots need to be induced and established prior to transfer to the nursery. This is an additional reason why ramets are prone to desiccation in the early stages of hardening. Shading can help to increase humidity and encourage good root growth. In practice, polythene tunnels/frames and shading (netting) are often used to construct a shade house (or rametry) (Fig. 3.12).

Fig. 3.12. a) Shade house from net; b) Shade house from polythene cover.

Oil palm ramets are first transferred from the culture medium to small bags filled with a mix of 50% soil and 50% sand. Ramets are then placed inside a transparent polythene frame within a double-layer shade house (netting) to maintain humidity for one to two months. One side of the shade house is also covered with a single layer of netting for the initial establishment of first stage. Afterwards, the polythene frame is removed but the ramets remain in the shade house. After a month, the side netting and one layer of roof shading (leaving one layer of roof shading) is removed for another one or two months. At the four- to six-frond stage, ramets can be transferred to the open main nursery and potted-on into large planting bags to allow further growth.

3.5   Off-site Facilities

Additional facilities needed to support the activities in the nursery include the following.

- **Incinerator.** Used to destroy the unused palms (diseased or infected palms, palms damaged by pests, abnormal palms from culling and palms left after an excess period of re-supplying.
- **Topsoil exploitation site.** A source of topsoil. This is normally situated outside, but close to, the nursery (See Sections 4.2.2 and 5.2.2).
- **Go-down/storage.** Needed to keep the tools and materials in the nursery. Chemical and non-chemical materials should be kept in different go-down/storage units. There must be warning signs and chemical use notices that are visible on the walls of chemical go-downs.

References

Bilck, A.P., Olivato, J.B., Yamashita, F., and de Souza, J.R.P. (2014) Biodegradable bags for the production of plant seedlings. *Polimeros.* 24(5): 547–553.

Chee, K.H., Chiu, S.B. and Chan, S.M. (1997) Pre-nursery seedlings grown on pot trays. *The Planter* 73(855): 295–299.

CIRAD (2008) *Germinated Oil Palm Seeds: Recommendations for Prenursery and Nursery Management.* Available at: https://www.palmelit.com/en/content/download/4239/32933/version/8/file/Booklet-of-recommandations-for-prenursery-and-nursery-management-oil-palm-seeds-CIRAD.pdf (accessed 30 October 2018).

Corley, R.H.V. and Tinker, P.B. (2015) *The Oil Palm,* 5th edn. Wiley Blackwell, UK, p. 639, ISBN: 978-1-405-18939-2.

Duckett, J.E. (1999) *A Guide to Oil Palm Nurseries.* The Incorporated Society of Planters, Malaysia.

Meadows, P. and Litz, V. (2016) *How to Manage an Oil Palm Nursery: A Guide for Smallholder Farmers.* Available at: Tetra Tech ARD, People, Rules, and Organizations Supporting the Protection of Ecosystem Resources (PROSPER) Liberia, https://land-links.org/wp-content/uploads/2018/04/USAID_Land_Tenure_PROSPER_How_to_Manage_An_Oil_Palm_Nursery.pdf (accessed 13 February 2019).

Muhamad, H., Subramaniam, V., Hashim, Z., Khairuddin, N.S.K. and May, C.Y. (2014) Water footprint: Part 1 – production of oil palm seedlings in peninsular Malaysia. *Journal of Oil Palm Research* 26(4), 273–281.

Muriuki, J.K., Kuria, A.W., Muthuri, C.W., Mukuralinda, A., Simons, A.J. and Jamnadass, R.H. (2013) Testing Biodegradable Seedling Containers as an

Alternative for Polythene Tubes in Tropical Small-Scale Tree Nurseries. *Small-scale Forestry* 13(2): 127–142.

Mutert, E., Esquivez, A.S., Santos, A.O. and Cervantes, E.O. (1999) The oil palm nursery: foundation for high production. *Better Crop International* 13(1), 19–44.

Noor, M.M. (2003) Zero burning techniques in oil palm cultivation: an economic perspective. *Oil Palm Industry Economic Journal* 3(1), 16–24.

Nordiana, A.A., Wahid O., Esnan, A.G., Zaki, A., Tarmizi, A.M., Zulkifli H. and Norman K. (2013) Land evaluation for oil palm cultivation using geospatial information technologies. *Oil Palm Bulletin* 67, 17–29.

Sitepu, B., Setiawati, U., Nur, F., Laksono, N.D., Anwar, Y. *et al.* (2019) *Field Trials in Oil Palm Breeding: A Manual. Techniques in Plantation Science.* Forster, B.P. and Caligari, P.D.S. (eds). CAB International, Wallingford, UK, in press.

Tolan Tiga Indonesia (2011) *Nursery. Tolan Tiga Indonesia Standard Operational Procedures.* Unpublished.

Yusuf, S.A., Muhammad, I.A.Q. and Justina, O.L. (2014) Determinants of risk and uncertainty in oil palm nursery. *Journal of Economics and Sustainable Development* 5(11), 174–186.

Pre-nursery Activities

4

Abstract

The pre-nursery is a key part of the double-stage nursery and is the major difference between single- and double-stage nurseries. The activities involve the preparation of small planting bags, which are close-packed (no spacing). Thus, the area needed is minimal. Abnormal palms are easily observed and discarded, so only good quality palms are grown on. This chapter describes the bed preparation for small planting bags, the layout to assist watering and fertiliser treatments, how to avoid flooding, how to fill the planting bags with soil, pre-treatments of seeds and ramets, how to sow germinated seed and plant ramets, how to provide and construct shading and, once established, how to pot-on into large planting bags. The tools, equipment and materials needed are also listed.

4.1   Materials and Tools

Materials that are needed for pre-nursery activities are as follows:

- Planting bags (which, if at all possible, should be biodegradable), size 18 cm x 21 cm x 0.1 mm. The colour of planting bag is normally black.
- A pot tray system can be used to replace polybag usage with size 69 mm × 60 mm × 120 mm or 58 mm × 58 mm × 120 mm.
- Steel wire (22 m) or wood plank for installation of bed edge.
- Topsoil.
- Bamboo and shad netting (or oil palm fronds) for shade construction.
- Water.
- Germinated seeds or ramets.

The tools needed:

- Planting bag labels.
- Wooden stick for making holes in the planting bag for drainage.
- Watering can for watering before and after planting.
- Sifter and hoe for filling planting bags with soil.

4.2 Preparation

4.2.1 Bed preparation for sowing germinated seed

The basic requirements in siting the beds are the following:

- Location of the bed(s) should be a part of, or adjacent to, the main nursery to allow for ease in palm transfer and reduce palm damage and stress during palm transfer.
- The area should be clean and clear of any debris such as roots, branches, leaves and weeds to reduce pests and diseases.
- The area should be flat and well drained (as described in Chapter 3) to avoid flooding during rainy seasons.

Methods used for making the beds are these:

- Beds are made lengthwise in an east–west direction.
- The width of a bed is normally 1.2 m and length 20 m but can be variable depending on needs. This design can accommodate 2,000 planting bags. Duckett (1999) describes a bed 1.22 m wide and 45.7 m long, that can accommodate 6,335 small planting bags.
- The distance between beds is normally 0.6 m.
- The dimension of the beds (width, length and spacing) is designed to provide easy access for workers from both sides so that they are able to reach all the planting bags.
- The base of the beds should be raised about 5 cm to prevent puddles of rainwater pooling in the beds.
- All the edges of the beds are fitted with steel wire or wooden planking (with dimensions that follow the shape of the beds) to keep the bags tidy and in position. Every corner or edge of the beds is set with a wood stick to retain the border planks (Fig. 4.1).

Fig. 4.1. Pre-nursery beds.

4.2.2 Filling planting bags

The soil media is a vital component for strong seedling and ramet growth. Care is needed in the proper filling of planting bags.

- The soil medium should be a mineral topsoil (not peat), and free of plant debris (roots, twigs, leaves, etc.), trash, stones and gravel.
- The soil should be sandy, clay loam, which has good drainage properties. For better texture, the topsoil can be mixed with compost (e.g. decomposed empty fruit bunches) with a compost–soil ratio of 1:3.
- For topsoils with a high clay content, the soil may be mixed with sand at a ratio of 1:3. Compost may also be added to the soil mixture.
- Fertiliser (rock phosphate) is normally added at a ratio 1:4.5, or can be added individually at 10 g per planting bag (Duckett, 1999).

Methods for filling planting bags with soil medium.

- Sieving the soil helps to remove plant debris (roots, twigs, leaves, etc.) as well as trash, stones, and gravel.
- Planting bags (23cm x 15cm x 0.1 mm) are filled with compacted top soil (± 1 kg/planting bag). There should be no empty space in the planting bag.
- Place filled planting bags correctly in rows in the prepared beds.
- Workers are often paid at piece rates, therefore it is important to calculate the amount of work needed. A worker can fill around 1,000 bags per day.
- The prepared planting bags should be set up in the beds a week prior to the seed/ramet arrival.
- The filled planting bags are watered until they are wet but not flooded.
- If necessary, planting bags may be topped up with additional soil medium.
- Place the 24 or 32 cell tray on a flat floor.
- Fill up the tray with planting medium and level off.
- Lift up the tray to a height of about 0.5 m and drop it on the floor. This compacts the planting medium. Repeat this step 2 - 3 times to make sure the medium is compacted.
- Level the medium off and place on a bench, it is now ready for sowing (see Fig. 3.4).

4.2.3 Shade construction

Shading (Fig. 4.2) is recommended in certain situations (see below). Although shading is strongly recommended by Turner and Gillbanks (1974) and is used as standard practice (Duckett, 1999), modern methods in irrigation techniques (see Chapters 3 and 7) have made shading redundant in many situations.

- Shading is recommended when bare root seedlings or ramets are being planted (ramets are normally shaded in a more specialised rametry, see Chapter 3).

Fig. 4.2. a) Shading by netting; b) Shading by oil palm fronds.

- Shading is recommended when watering procedures are inadequate or suspect.
- Shading is recommended as a safeguard when growing valuable, unusual or exotic germplasm. For example, breeding materials such as the South American oil palm (*Elaeis oleifera*), hybrids between *Elaeis guineensis* and *E. oleifera* or mutant palm (see Setiawati *et al.*, 2018, this series) populations (see Nur *et al.*, 2018, this series).

Methods in constructing a simple shade nursery area are as follows:

- The sides of the nursery shade area should be 2 m high to provide air circulation and moderate sunlight penetration in the mornings and afternoons. The height also provides an easy (roomy) workspace for workers.
- The shade netting is supported by bamboo poles or rounded wooden rods with a diameter of about 7.5 cm.
- The shade house is usually arranged in a 3 x 3 x 3 m cube design.
- The roof of the shade house can be made up of oil palm fronds (or nipah palm fronds, or other such locally available material) with a shading rate of up to 80%. **Note:** the fronds should be sprayed with a herbicide and pesticide to ensure they are not a source of pests and diseases.
- The shade should be reduced gradually (see below), i.e. during the month prior to palm transfer to the (open) main nursery as follows:
  - the first month reduced to 30%;
  - the second month reduced to 25%; and
  - the third month the roof should be fully open.

4.3 Practices for Planting Germinated Seeds and Ramets

- A field assistant should be assigned to supervise the planting.
- Planting of germinated seeds should be completed as soon as possible, preferably no later than a day after arrival.

- Training is required as the work is delicate. On average, one worker can plant between 5,000 and 7,000 seeds per working day.
- Care must be taken to ensure that the worker can differentiate between the radicle (root) and plumule (shoot); if these are inverted during planting the resulting palm will be abnormal.

4.3.1 Steps in planting germinated seed

Methods in planting germinated seeds into small planting bags and into pot trays are similar. Steps in planting germinated seeds are following:

Step 1
Select the normal germinated seed. Off-type germinated seeds are shown in Fig. 4.3 and should be discarded for commercial seed production (Kelanaputra *et al.*, 2018).

| Globular | Atrophied | Twisted | Twisted | Longer Root | Longer Root |

Fig. 4.3. Off-type germinated seeds.

Step 2
Count out the number of seeds that are needed for planting.

Fig. 4.4. Count out the seeds.

Step 3
Transfer the germinated seeds (e.g. from a reputable seed supplier) to the nursery. This should be carefully controlled, and the number of seeds received should correspond to that ordered. The seed is formally handed over to the nursery supervisor.

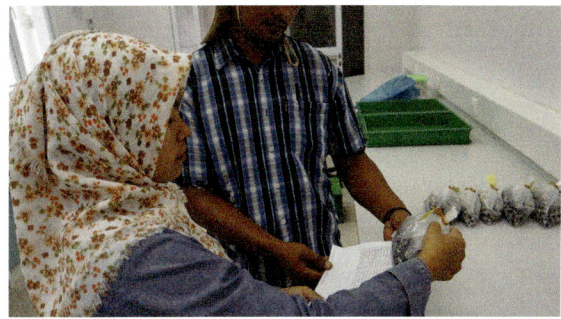

Fig. 4.5. Handover of germinated seed from seed supplier to the nursery supervisor.

Step 4
Before planting the germinated seed, beds should be watered until they are wet but not flooded.

Fig. 4.6. Watering the beds of bags.

Step 5
Distribute the seeds to workers based on the planting plan.

Fig. 4.7. Group the germinated seeds in the bed.

Step 6
Germinated seeds are soaked in tap water prior to planting, so that they are fully imbibed. This activity is conducted simultaneously during seed laying (next step).

Fig. 4.8. Soaking the germinated seeds.

Step 7
Planting is normally carried out with two to four workers working simultaneously within a bed. The germinated seeds are laid individually on the soil surface of the planting bags or tray compartment before planting.

Fig. 4.9. The germinated seed placed on top of individual bags in the bed.

Step 8
Holing and planting the germinated seeds is carried out simultaneously. The depth of hole should be 2–3 cm. The radicle should be pointing *downwards* (Fig. 4.11).

Fig. 4.10. Holing and planting.

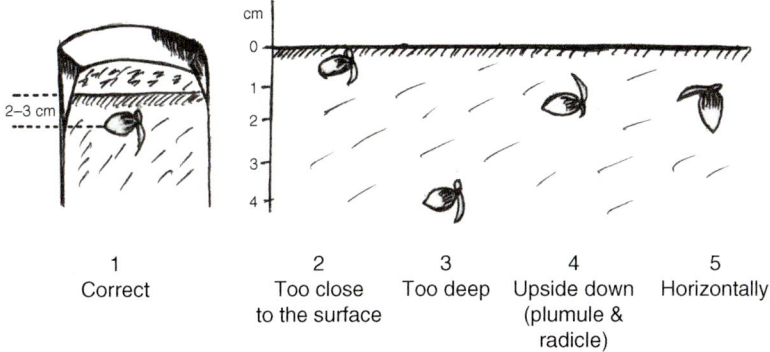

Fig. 4.11. Correct and incorrect orientation of germinated seed.

Step 9

Install a signboard in the bed as soon as seed planting is complete.

Fig. 4.12. Label the planted seed at the edge of bed.

Step 10

The planted seeds should be watered as soon as planting is complete. The soil in the bag should be wet but not flooded.

Fig. 4.13. Watering after planting.

4.3.2   Planting of clonal plantlets (ramets)

Ramets or clonal plantlets are planting materials propagated from tissue culture and have the genetic make-up (clones) of their parental ortet. Ramets that are ready to be planted are normally received as bare-rooted plantlets at the two- to three-leaf stage (Tan, 2011).

Steps in planting the ramets: Step 1
Soil medium and planting arrangements are the same as for germinated seeds (see above).

Fig. 4.14. Ramet ready to planted.

Fig. 4.15. Ramets inside a rametry.

Step 2
Make the planting holes using a wooden stick.

Step 3
The depths of holes should be sufficient to accommodate the roots.

Step 4
The roots are placed into the hole and covered with soil up to the collar of the bag or tray compartment. No root should be exposed on the soil surface.

Step 5
Ramets usually bear a field code which consists of a clone code and origin. This information should be transferred to the planting bag/tray label, once planting is completed.

4.3.3   Transfer to main nursery and potting-on

Pre-nursery seedlings and ramets are transferred to the main (open) nursery when they have reached the four-leaf stage of development (10 to 14 weeks after planting). The first culling of off-type plants is carried out during the transplanting stage; this activity is discussed in more detail in Chapter 5.

The pot tray system has a higher capacity for seedlings than small poly-bags. Up to 1,400 seedlings can be transported by a 3 ton-lorry.

4.4   Steps in Setting up a Pre-nursery for Subsequent Field Trialling

Pre-nursery activities for field trial work are similar to commercial palm production methods, but with some differences.

Step 1: Design of beds
Breeding trials use various planting materials or progenies (see Sitepu *et al.*, 2019, this series). The number of palms per progeny or clone may vary, but

200 seedlings/ramets is common. The number is usually in excess of what is required for field planting, as it may be reduced due to culling and other losses, both in the nursery and in the field (plant losses in the field can be compensated by re-planting with extra 'supply' palms). Some 200 germinated seeds will require a bed of 10 planting bags wide, and 20 planting bags long (Fig. 4.16).

Fig. 4.16. Design of bed for a typical breeding population for trialling.

Step 2: Labelling

Breeding trials often involve a variety of planting materials. There should be one bed per progeny or clone to avoid mixing between planting materials. A label should be attached to each planting bag and should identify each seed or ramet. Be careful to match labels when planting seeds, potting-on and field planting. It is strongly recommended that seeds be handled by batch/bed group separately.

Fig. 4.17. Matching the label between germinated seed and bag.

Step 3: Potting-on

To avoiding mixing up progeny and clones, transfer palms in their planting bags to crates or baskets one bed at a time. Seedlings/ramets are then transferred to the main nursery using crates or baskets consisting of only one progeny/clone each.

Fig. 4.18. Transfer to the main nursery.

References

Duckett, J.E. (1999) *A Guide to Oil Palm Nurseries*. The Incorporated Society of Planters, Malaysia.

Kelanaputra, E.S., Nelson, S.P.C., Setiawati, U., Sitepu, B., Nur, F., Forster, B.P. and Purba, A.R. (2018) *Seed Production in Oil Palm: A Manual. Techniques in Plantation Science*. Forster, B.P. and Caligari, P.D.S. (eds). CAB International, Wallingford, UK, p. 59.

Nur, F., Forster, B.P., Osei, S.A., Amiteye, S., Ciomas, J., Hoeman, S. and Jankuloski, L. (2018) *Mutation Breeding in Oil Palm: A Manual. Techniques in Plantation Science*. Forster, B.P. and Caligari, P.D.S. (eds). CAB International, Wallingford, UK, p. 63.

Setiawati, U., Sitepu, B., Nur, F., Forster, B.P. and Dery, S. (2018) *Crossing in Oil Palm: A Manual. Techniques in Plantation Science*. Forster, B.P. and Caligari, P.D.S. (eds). CAB International, Wallingford, UK, p. 96.

Sitepu, B., Setiawati, U., Nur, F., Laksono, N.D., Anwar, Y. *et al.* (2019) *Field Trials in Oil Palm Breeding: A Manual. Techniques in Plantation Science*. Forster, B.P. and Caligari, P.D.S. (eds). CAB International, Wallingford, UK, in press.

Tan, C.C. (2011) Nursery practice for production of superior oil palm planting material. In: *Agronomic Principles and Practices of Oil Palm Cultivation*. Goh, K.J., Chiu, S.B. and Paramananthan, S. (eds). Agricultural Crop Trust, Petaling Jaya, pp. 145–169.

Turner, P.D. and Gillbanks, R.A. (1974) *Oil Palm Cultivation and Management*. Incorporated Society of Planters, Kuala Lumpur, 672 ff.

Main Nursery Activities 5

Abstract

Two to three months after palm establishment in the pre-nursery (Chapter 4), the palms are potted-on into large planting bags in the main nursery. At this stage the seedlings (or ramets) need to be moved and given more space as they begin to shade each other out in the close-packed planting bag system of the pre-nursery (or rametry). The smaller planting bags also inhibit growth and development. Transplanting into large bags with a fresh soil medium and spacing is the first activitiy in the main nursery. Prior to potting-on in the main nursery, off-type seedlings and ramets are culled, thus only strong, uniform, healthy palms are advanced. Methods of transport to the main nursery, potting-on and spacing are described along with general husbandry of seedlings and ramets until they are ready for field planting.

5.1   Materials and Tools

Materials needed:

- Large planting bags (preferably biodegradeable): size 38 cm × 55 cm × 0.15 mm. The colour of planting bags is normally black.
- A good supply of mineral topsoil.
- Water.
- Compost.
- Fertiliser.
- Oil palm seedlings or ramets at the four-leaf stage (from the pre-nursery or rametry, respectively).

Tools needed;

- Steel wire with a diameter 22 mm and wood-stick for spacing.
- Line marking from wood (every 0.9 m has a mark) to facilitate spacing.

- Labels for planting bag/palm identity.
- Watering can for watering before and after planting.
- Soil sieve and hoe to fill the large planting bags.
- Tool to form planting hole.
- Signboard.
- Wheelbarrow to transport seedlings and ramets.
- Cutter or knife to remove seedlings or ramets from their small planting bags.

5.2 Preparation

5.2.1 Nursery area

The area of the main-nursery is proportional to the amount of young plants (seedlings and/or ramets) to be planted and grown on. One ha of nursery space can accommodate 12,000 young plants including the necessary planting bag-spaced plots and road access. Site preparation for main-nursery from levelling the area, arranging nursery drainage and constructing irrigation should be completed two months prior to palm transfer from the pre-nursery (see Chapters 3 and 4).

5.2.2 Filling large planting bags with soil medium

Filling planting bags with sieved soil is a simple task. The work is usually carried out by piece-workers or at contract rates. The work of filling the planting bags should begin at least a month prior to the established four-leaf seedlings or ramets being transferred to the main nursery for planting out, and should be completed a week prior to potting-on.

The size of planting bag may vary (see various recommendations in Table 5.1). Here we recommend a size of 38 cm × 55 cm × 0.15 mm. The planting bag should be twice the volume of the small pre-nursery planting bag. Piece-workers can normally fill 180 bags in a working day, but two workers taking turns to fill and place filled bags can complete 500 or more bags per day (Duckett, 1999).

Table 5.1. Various recommendations for planting bag size.

Age of palm (months)	Rankine and Fairhurst (1998)	Duckett (1999)	Turner and Gillbanks (2003)	Mathews *et al.* (2010)
03–12	40 × 45 cm	38 × 45 cm	40 × 50 cm	41 × 46 cm
12–18	40 × 45 cm	45 × 60 cm	45 × 60 cm	41 × 46 cm
18–24	40 × 45 cm	60 × 75 cm	45 × 60 cm	41 × 46 cm

Precautions in preparing soil medium

- The area that the soil is taken from should be free of *Ganoderma* (a major disease of oil palm) and should not have been treated recently with herbicides or chemicals. If treated, time should be given for the toxic residues to break down (such toxins can be disastrous for young oil palm plants, particularly where hormone-type herbicides are involved).
- Soil should not be collected from areas that have been burnt; such soil can be detrimental to the palms.
- The best soil types should be used, i.e. mineral topsoil with a good soil texture/mixture of sandy-clay loam and with a good water absorption capacity.
- Soil mixing can be done to improve soil texture, e.g. a high clay soil should be mixed with sand in a ratio of 1:3 (sand:soil); compost (decomposed empty fruit bunches from an oil palm mill) may also be added in a ratio 1:3 (compost:soil).
- Peat soil is not recommended, but when there is no option the peat soil should be mixed with mineral soil at a ratio of 1:3 (soil:peat).
- The soil medium should be free from large clods, stones, plant debris, wood, gravel and other detritus.
- Wet soil, especially soil with a high clay content, will lead to soil compaction which inhibits root development. Therefore, it is better to prepare the soil medium in dry weather conditions, if possible.

Steps in filling planting bags with soil: Step 1
Select the site to collect the topsoil with consideration of the points above.

Step 2
Clean the soil by sieving to separate the debris.

Step 3
Fill the planting bag with approximately 20 kg soil/planting bag. There should be less than a 2.5 cm gap at the top of the planting bag.

Step 4
The planting bag is compacted by bumping on the ground to consolidate the soil, not by pressing the soil.

Step 5
The filled bags are then sent to the block prior to spacing using a wheelbarrow.

5.2.3 Spacing the planting bags

The planting bags are spaced in the nursery to provide good lighting and reduce competition among neighbours throughout the period in the main nursery, i.e. up to field transfer. The spacing also allows workers to pass freely among the palms for nursery observation and maintenance. The spacing allows the layout of the irrigation system (see Chapter 3). Once in position the planting bags remain in place until transfer to the field.

Steps in planting bag spacing: Step 1
Four workers are normally required for spacing: two for lining and two to arrange the planting bags.

Step 2
Determine the line heads by using steel wires.

Step 3
Pull the second steel wire in the opposite direction to be lined, to form a row.

Step 4
Fit with line marking (parallel to road) for planting bag arranging points sized 0.9 m × 0.9 m (Fig. 5.1).

Step 5
Transfer the planting bag by wheelbarrow and put the planting bag at the arranged point. Carry the planting bag carefully; avoid holding the lips of the bag to avoid splitting.

Step 6
The planting bags are carefully placed upright (to avoid tearing).

Step 7
The lining and arrangement of planting bag should be completed at least two weeks prior to potting-on.

Fig. 5.1. Spacing the bags using steel wire and line marking.

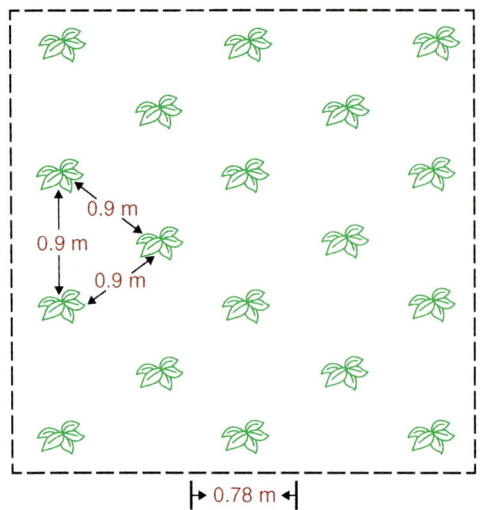

Fig. 5.2. Triangular layout of bag spacing.

5.2.4   Holing

Holing is carried out prior to potting-on the seedlings/ramets in the main nursery. The hole should be the same dimensions as the small planting bag (10 cm diameter). These holes are prepared using a planting corer (Fig. 5.3). Regular cleaning of the corer should be performed to prevent the soil medium from sticking to it and thus reduce disease incidence. A pieceworker can core about 600 holes per day with this tool. The removed core should be placed next to the bags for supplying soil subsequent to transplanting (see Figs 5.4, 5.5).

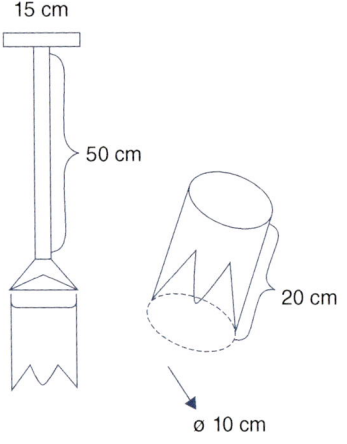

Fig. 5.3. Design of planting hole corer.

Fig. 5.4. Form of hole.

Fig. 5.5. Workers holing the bags.

5.3 Practices

5.3.1 Transplanting to the main nursery

The young plants (seedling or ramets) are removed from the pre-nursery after about two to three months, or when four to five leaves have emerged (see Chapter 4). Roots will begin to emerge from the drainage holes in the planting bags after two months, thus these emerging roots will be disturbed during the potting-on process. Culling of abnormal plants should be carried out at this time (Fig 5.6 shows examples of off-type plants that are culled). One worker can normally pot-on 500 plants per day, but the work will go faster if it is done by a group of four or five workers with devolved tasks (opening bags, placing plants in the holes and compacting the soil).

The pot tray system is simple and speedy for transport to and transplanting in the nursery and it is possible to transfer up to 1,000 plants per day (Chee, 1997).

Steps in potting-on palms from pot trays to large planting bags: Step 1
Extract the young seedlings from pot tray.

Step 2
Poke a hole in the planting medium in the large planting bag, apply fertilizer according to standard policy.

Step 3
Place the seedling in the hole and gently compact the medium to remove any air pockets.

Fig. 5.6. Examples of abnormal seedlings: a) Narrow leaf; b) Crinkled leaf; c) Twisted leaf; d) Rolled leaf.

Steps in potting-on seedlings from small to the large planting bags: Step 1
Transfer young plants (seedlings or ramets) from the pre-nursery to the block of the main nursery. Use crates or boxes to transfer progenies (or clones) separately to avoid any chance of mixing different planting materials.

Step 2
Water the planting bag prior to potting-on.

Step 3
Doubleton, tripleton and tetraton palms (seeds that have produced two, three or four shoots/palms respectively) should be separated at this point and planted separately into large planting bags.

Step 4
The young plants (seedlings or ramets) are placed out next to their new planting bags.

Fig. 5.7. Place young plants next to their new bags.

Step 5
Remove the small planting bag using a cutter or scissors, taking care to keep the soil intact.

Fig. 5.8. Removing the seedling from the small bag.

Step 6
The young plants are placed into the prepared holes (the bole should remain on the surface) and the soil medium compacted around each one by hand, ensuring that the planting is level with the pre-nursery soil level.

Fig. 5.9. Compacting the soil by hand.

Step 7
Watering should be done immediately after planting.

Step 8
After transplanting, label the plants using a signboard placed in front of each group of seedlings/ramets.

5.3.2 Supply palms

Supply palms are additional young plants of the same age that may be needed to replace those that are substandard, die or have been damaged subsequent to field planting. A number of supply palms may be kept in the main nursery for one year after planting the same material.

Supply palms (palms remaining in the nursery after the main batch has been taken to the field) are often scattered across the nursery. These are then collected and taken to a designated area. The spacing used depends on the period for which the palms are to be maintained in the nursery. The spacing of the bags determines the number of palms:

- 1.5 m triangular gives 4,972 bags per hectare;
- 1.8 m triangular gives 3,455 bags per hectare;
- 2.1 m triangular gives 2,538 bags per hectare; and
- 2.4 m triangular gives 1,942 bags per hectare.

Fig. 5.10. Supply palms maintained in the main nursery.

During the holding period, the supply palms can gain a height of more than 1.5 m. This will lead to etiolation and elongation and therefore topping or pruning back to 1.5 m height is advisable. Cutting back is facilitated using a stick 1.5 m long. The stick is aligned to one frond, which is then cut along with the next youngest frond (Figs 5.11, 5.12).

Fig. 5.11. Topping supply palms.

Fig. 5.12. Topped supply palms.

5.3.3 Transport to the field

Young palms can be planted in the field 9–12 months after arrival in the pre-nursery (6–9 months in the main nursery). At this stage, the palms have a well-developed root system, a full natural frond spread, good stature and a good state of nutrition.

Field-ready young palms are normally carried to the field by a tractor with trailer (container capacity of 100 to 110 planting bags), placed upright in a single layer. Double layers are possible, but require extra care as the fronds of the lower layer are easily disturbed and damaged. Loading and unloading double-layer load trailers also incurs a time delay, which may be significant when transporting large numbers of palms to more distant planting areas.

Fig. 5.13. Transportation of young palms to the field.

5.4 Plant Production for Field Performance Trials

The management practices in the main nursery for the production of commercial and trial planting material are quite similar. The main difference is that care needs to be taken to prevent mixing different genetic materials for trialling. This is not a major issue for commercial planting as here the material is usually of the same origin. Thus, the difference between trial and commercial management is that in trials it is important to treat each progeny or clonal batch separately.

5.4.1 Placement of plants for trial work

As discussed in Chapter 4, normally about 200 seedlings/ramets are used for field performance testing (trialling). After the first culling prior to potting-on in the main nursery the number of seedlings/clones transplanted will be reduced by about 0–10% (180–200 palms remaining). Thus, an excess is normally grown on in the nursery. There is also potential for planting bags to be mixed. Thus, labelling is essential: each planting bag should be labelled and the arrangement of different progenies or clones should be considered to reduce mix-ups.

5.4.2 Steps in planting bag placement for subsequent trial work: Step 1

A block in the main nursery should be assigned specifically for materials for subsequent trialling, and the breeder concerned notified.

Step 2
A group of planting materials should comprise a minimum of 170 palms. Thus the dimensions of the main nursery block need to be considered. The length of blocks for trial work are commonly 50–100 m.

Step 3
Calculate the capacity of planting bags per row. A 50 m block will accommodate 50 seedlings/clones, and a 100 m block 100 seedlings/clones.

Step 4
Determine the number of rows needed for each planting material. A 50 m block length will require four rows for each planting material, and 100 m block length will need two rows for each planting material.

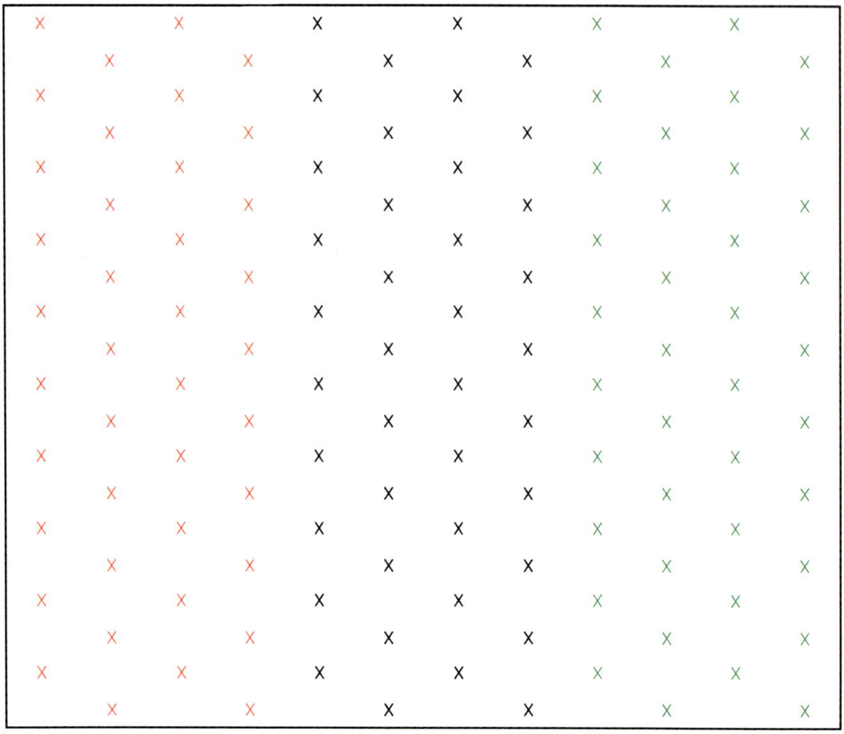

Fig. 5.14. An example of bag placement for different planting materials (red, black and green) within a main nursery block used to rear young plants for field trialling.

Step 5
Record the location and row for each planting material.

5.4.3   Supply palms for trial work

Oil palm trials often use 64 palms which are planted in four replicate blocks (in a 4 × 4 design, see Sitepu *et al.*, 2019, this series). After a year in the main nursery, and after three rounds of culling, the number of palms available for field planting is reduced. Thus an excess number of young palms is needed. Furthermore, an excess is also required as supply palms because some palms may need to be replaced after planting in the field. Replacements are normally maintained in the nursery for an additional 12 months.

5.4.4   Transport to the field for trial work

Transporting field-ready palms from the main nursery to the field needs care and attention as mix-ups are easily made. Care is needed in maintaining individual progenies or clones together during transport with block randomisation being done at the trial site. This work is complicated and needs to be carefully supervised. It is described in detail by Sitepu *et al.* (2019, this series).

Steps in transporting young palms in bags to the field for trial work:
Step 1
All steps should be undertaken in consultation with the person responsible for the trial work.

Consider the field planting layout prior to transplanting the palms. Commonly, a field trial plot consists of 8 or 16 palms. A 16-palm plot is considered here.

Step 2
The easiest system of dropping off the planting bags from the tractor-trailer to the field is by taking a snaking pathway, A-B-C-D-E, etc. (Fig. 5.15).

Step 3
Organise the sequence of seedlings/clones to correspond to the snaking pathway. The planting materials placed first in the trailer will be the last ones dropped off (F plants) and the last to be placed in the trailer will be the first to be dropped off (A plants).

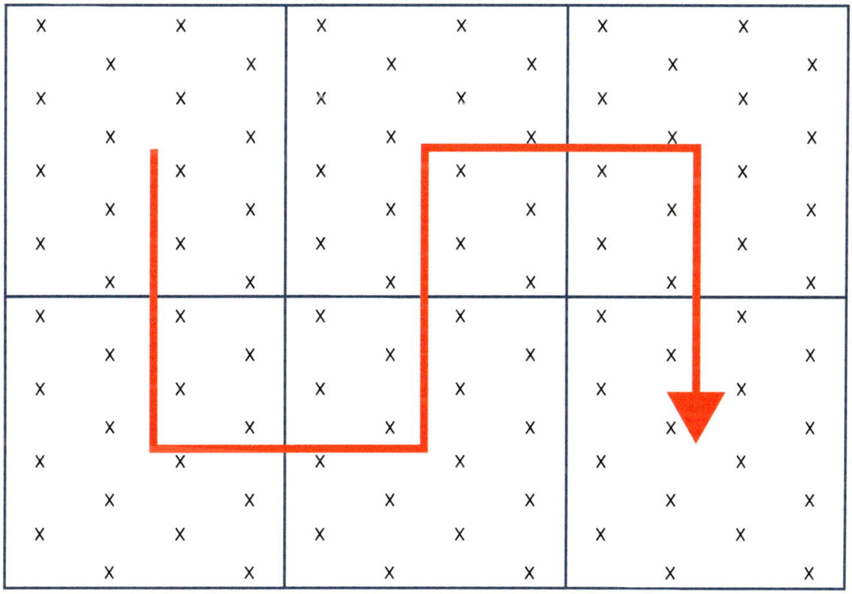

Fig. 5.15. Arrangements of transporting the bags to the field planting.

References

Chee, K.H., Chiu, S.B. and Chan, S.M. (1997) Pre-nursery seedlings grown on pot trays. *The Planters* 73(855): 295–299.

Duckett, J.E. (1999) *A Guide to Oil Palm Nurseries*. The Incorporated Society of Planters, Malaysia.

Mathews, J., Tan, T.H., Yong, K.K., Chong, K.M., Ng, S.K. and Ip, W.M. (2010) Managing oil palm nursery: IOI's experience. *The Planters* 86(1016): 771–785.

Rankine, I. and Fairhurst, T. (1998) *Oil Palm Nursery – Field Handbooks*. Potash and Phospate Institute, Singapore.

Sitepu, B., Setiawati, U., Nur, F., Laksono, N.D., Anwar, Y. *et al.* (2019) *Field Trials in Oil palm Breeding: A Manual. Techniques in Plantation Science.* Forster, B.P. and Caligari, P.D.S. (eds). CAB International, Wallingford, UK, in press.

Turner, P.D. and Gillbanks, R.A. (2003) *Oil Palm Cultivation and Management*, 2nd edn. The Incorporated Society of Planters, Malaysia.

Fertiliser Programmes

6

Abstract

Fertiliser application is a major activity in the nursery. The materials and types of fertiliser for nursery palm plants vary. Fertiliser recommendations can be different among oil palm nurseries in different companies due to varying management practices and soil types used. It is recommended to seek the opinion and recommendations of an agronomist when setting up a nursery and when problems arise. Detailed recommendations of fertiliser applications in oil palm nurseries have been described and altered over time (see Rankine and Fairhurst, 1998; Duckett, 1999; Turner and Gillbanks, 2003; Mathews *et al.*, 2010; Tan, 2011). The correct fertiliser type and dose are important not only for maximising plant health, but also to minimise costs as fertilisers are an expensive nursery input. Fertiliser application aims to fulfil the nutrient requirements of young plants. Best agronomy practices will help reduce fertiliser wastage by evaporation, leaching, overdose, etc. This chapter describes guidelines and methods in fertiliser application.

6.1 Considerations for Fertiliser Programmes

- Solid fertilisers should be applied approximately 5–7 cm away from the palm base. Care must be taken to avoid contact with the young leaves either directly (fertiliser particles) or indirectly (worker's gloves).
- Fertiliser can cause plant scorching if not applied correctly. Excessive fertiliser application is unlikely to have any advantage for plant growth, and is more likely to have adverse effects. Thus, proper measures and correct fertiliser usage are important. Correct rates of fertilisers should be provided to avoid irregular or surplus applications that are not only wasteful but may prove toxic to young plants.
- An agronomist's involvement in preparing a fertiliser programme is desirable to determine optimal dose treatments.

- Store the fertiliser safely in the store.
- Use personal protective equipment (PPE) to minimise the risk of inhalation, eye irritation and contact with open wounds/damaged skin (see Chapter 2).
- Provide clean water for hand and face washing after work with fertiliser.

Fig. 6.1. a) Worker wearing complete PPE; b) Measured cup for fertilising; c) Granular fertiliser direct application; d) Fertiliser on the bag.

6.2 Fertiliser Programmes for Pre-nursery

Fertilisation in the pre-nursery and rametry involves liquid applications (mixing fertiliser with water) and therefore a soluble fertiliser should be used. The application of fertiliser can be integrated with irrigation (fertigation).

Fertigation is a modern agro-technique, which provides an excellent opportunity to maximise palm growth and development, and minimise environmental pollution by increasing fertiliser use efficiency, reducing fertiliser rates and maximising the return on fertiliser cost. In fertigation, timing, amounts and concentration of fertilisers applied are easily controlled. There is no risk of foliar scorch and subsequent development of plant pathogens. Fertigation has beneficial effects on palm height, stem height, stem diameter, number of leaves and number of leaflets, leaf length, leaf width and chlorophyll index (Rahmana *et al.*, 2017).

6.2.1 Considerations in applying fertiliser to pre-nursery and rametry plants

- One hour after applying a manure solution to month-old seedlings or ramets, the plants should be watered.
- At two to three months old, plant fertilisation application is carried out after watering to reduce leaching.
- Fertiliser application should be done carefully, avoiding contact with leaves.
- Apply liquid compound fertiliser to leaves when deficiency symptoms appear, but be careful to apply the precise dose (Tables 6.1, 6.2).

Table 6.1. Dose rates for fertiliser application at different plant ages (pre-nursery or rametry) (Tolan Tiga Indonesia, 2011).

Months after planting	NPK 15.15.6.4 (g)	Application
1	8 g/100 seeds	4x/month: 8 g/5L water for 100 plants
2	1 g	4x/month: 4 g/5L water for 100 plants
3	1 g	4x/month: 4 g/5L water for 100 plants

Table 6.2. Dose rates for fertiliser application with fertigation (Netafim, 2018).

Age of palms (weeks)	N	P_2O_5	K_2O	MgO
	Grams per plant			
1	0.8	0.8	0.3	0.2
3	0.8	0.8	1.2	0.1
5	1.1	1.1	0.4	0.3
7	1.2	1.2	1.7	0.2

6.3 Fertiliser Programmes in the Main Nursery

Two fertiliser programmes are used in the main nursery. The first is the use of granular fertiliser, the second uses fertigation. Fertiliser dose is dependent on soil type and fertiliser applications should comply with recommendations issued by the company (or company's agronomist). Some recommended doses for plants in bags are listed in Table 6.3.

6.4 Fertigation

One of the most important challenges facing nursery growers and managers today is to provide oil palm with the optimal quantity of water and nutrients in the most cost-efficient manner possible. For intensive high-quality oil palm plant production, the best answer to this challenge is fertigation, where both water and fertilisers are delivered in small quantities, accurately and uniformly to each individual palm in each bag through emitters. In view of the limited soil volume in a bag, fertigation becomes essential to continuously feed the palms, thus matching plant needs according to developmental stage with an appropriate mix of nutrients to achieve optimal crop performance without any wastage. Fertigation, a modern agro-technique, provides an excellent opportunity to maximise palm growth and development, and minimise environmental pollution by increasing fertiliser use efficiency, reducing fertiliser rates and increasing return on the fertiliser invested. In fertigation, timing, amounts and concentration of fertilisers applied are easily controlled. There is no risk of foliar scorch.

Table 6.3. Dose rates for fertiliser application in main nursery (Duckett, 1999; Netafim, 2018).

Age of palms (weeks)	Granular (g/plant) (Duckett, 1999)		Fertigation (g/plant) (Netafim, 2018)			
	Quantity (g)	Type and ratio	N	P_2O_5	K_2O	MgO
9	3.5	NPKMg 15/15/6/4	1.5	1.5	0.6	0.4
10	3.5	NPKMg 15/15/6/4	1.5	1.5	0.6	0.4
12	7	NPKMg 15/15/6/4	1.2	1.2	1.7	0.2
14	7	NPKMg 12/12/1712 + TE	1.2	1.2	1.7	0.2
16	7	NPKMg 15/15/6/4	2.3	2.3	0.9	0.6
18	7	NPKMg 12/12/17/2 + TE + Kieserite	1.8	1.8	2.6	0.3
20	7	NPKMg 15/15/6/4	1.8	1.8	2.6	0.3
22	7	NPKMg 12/12/1712 + TE	3.0	3.0	3.0	3.0
24	7	NPKMg 15/15/6/4 + Kieserite	2.4	2.4	3.4	1.3
26	15	NPKMg 12/12/17/2 + TE	2.4	2.4	3.4	1.3
28	15	NPKMg 15/15/6/4	2.4	2.4	3.4	1.3
30	15	NPKMg 12112/17/2 + TE	2.4	2.4	3.4	1.3
32	15	NPKMg 15/15/6/4 + 15 g Kieserite	3.0	3.0	4.3	1.4
34	30	NPKMg 12/12/17/2 + TE	3.0	3.0	4.3	1.4
36	30	NPKMg 12/12/17/2 + TE	3.0	3.0	16.3	2.0
38	30	Kieserite	3.0	3.0	16.3	2.0
40	30	NPKMg 12/12/17/2 + TE	3.0	3.0	4.3	2.0
42	30	Kieserite	3.0	3.0	4.3	2.0
44	30	NPKMg 12/12/17/2 + TE	3.6	3.6	5.1	2.0
46	30	NPKMg 12/12/17/2 + TE	3.6	3.6	5.1	2.0
48	30	NPKMg 12/12/17/2 + TE	3.6	3.6	5.1	2.0
51	30	Kieserite	3.6	3.6	5.1	4.0
54	30	NPKMg 12/12/17/2 + TE	3.6	3.6	5.1	5.3
57	30	NPKMg 12/12/17/2 + TE	3.6	3.6	5.1	5.3
60	30	NPKMg 12/12/17/2 + TE + Kieserite	3.6	3.6	5.1	5.3

Irrigation water requirement is decreased by 36 to 52% and fertiliser application rates by 7.5 to 34.7% when compared to sprinkler irrigation and soil application of fertilisers (Netafim, 2018). Other benefits from fertigation include: reduced nursery period (8.5 months only); improved nutrient availability; enhanced plant nutrient uptake; prevention of nutrient losses by leaching and runoff; and decreased weed infestation. The benefits of drip irrigation with fertiliser are described in Section 7.1.1.

Netafim has technology for drip irrigation that can also be used for fertigation – the FertiKit™ 3G based on Netafim's unique Nutrigation™ technology. This technology controls the amount of water and fertilisers used, optimising resource utilisation for each specific oil palm plant, both in pre- and main nursery and soil/substrate. The FertiKit™ 3G is a fully con-figurable fertiliser/acid dosing unit – a highly cost-effective solution for pre-cise Nutrigation™. Based on a standard platform, the FertiKit™ 3G offers seven different operation modes, selectable according to the site conditions, in order to maximise usage of available water flow rate and pressure on the main irrigation line, ensuring the highest efficiency with minimum invest-ment. The FertiKit™ 3G doses the various fertilisers and acid into a homoge-neous solution and injects it into the irrigation water main line. The suction of the fertilisers and acid in the dosing channels is based on the Venturi principle. This requires a pressure differential – available on the main line or supplied by the main line pump or the FertiKit's dosing booster.

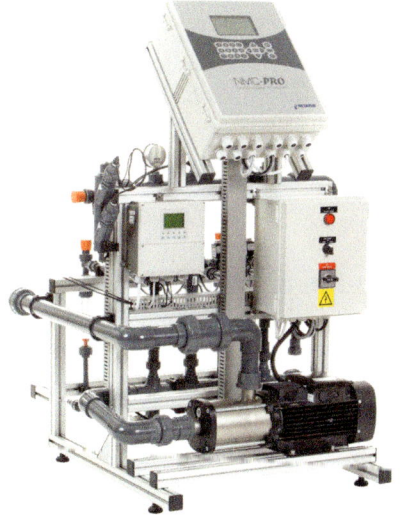

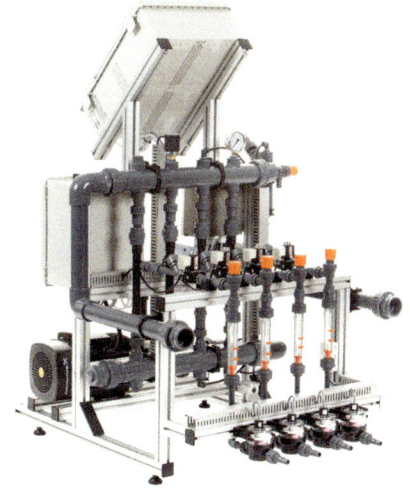

Fig. 6.2. FertiKit™ 3G Netafim technology.

A large range of water-soluble fertilisers, both solid and liquid, are suitable for fertigation depending on the physicochemical properties of the fertiliser solution. For example, urea is a cheaper source of N (per unit nutrient) because of its high N concentration, but its use has not been as widespread as ammonium sulphate because of high N losses, mainly through volatilisation. This is accentuated in the humid conditions maintained in oil palm nurseries. On the other hand, ammonium sulphate increases soil acidity, which has deleterious effects on plant growth. The main factors considered in selecting fertilisers for fertigation include: (i) plant type and stage of growth; (ii) soil conditions; (iii) form (solid or liquid); (iv) solubility and compatibility; (v) water quality; and (vi) fertiliser availability and price. The type of fertiliser for fertigation should be of high quality, with high solubility and purity, containing low salt levels and with an acceptable pH, and it must fit in with the nursery management programme. In Southeast Asia, the two most frequently used compound fertiliser formulas for oil palm plants are 15-15-6-4 (nitrogen (N)-P2O5-K2O-MgO)) and 12-12-17-2+micronutrients (N-P2O5-K2O-MgO+boron (B), zinc (Zn), manganese (Mn), etc.). Tables 6.2 and 6.3 provide a generic nursery fertiliser schedule which can assist nursery growers and consultants in calculating fertiliser rates based on the types and sources of materials available.

6.5   Care in the Use of Fertilisers in the Nursery

Nutrient and trace element deficiencies can occur in nurseries and cause setbacks to the young plants. In certain cases, these can arise from inherent inadequacies in the soil, either at the initial nursery stage or after several seasons of bag filling on site with depleted topsoil reserves, or from heavy irrigation that causes leaching of nutrients. Some common nutrient deficiency symptoms are listed below.

*   **Boron (B) deficiency** is characterised by young leaves becoming curly and crinkled. Corrected by foliar spraying with a solution of 10 g borate in 5 l water for 100 plants.
*   **Magnesium (Mg) deficiency** is characterised by a yellowish colour in lower leaves (older leaves). Corrected by foliar spraying with a solution of 20 g 'English salt' + 5 l water for 100 plants with an age of less than three months. Plants older than three months can be treated with Kieserite, a dose of 10 g (3–6-month-old plants) or 20 g (6–12-month-old plants).
*   **Nitrogen (N) deficiency** is characterised by a pale green colour in all leaves. Corrected by spray with a solution of 7 g urea + 5 l water for 100 plants with an age less than three months. Older plants may be sprayed with a solution of 14 g urea + 5 l water for 100 plants.

- **Phosphorus (P) deficiency** has no special foliar characteristic, but results in poor root development which affects growth and trunk diameter. Phosphorus deficiency should not appear if soil preparations are conducted correctly.
- **Potassium (K) deficiency** is characterised by short pinnae, pale green spots that turn to a translucent orange. The palms may develop a 'flat-top' appearance. Corrected by replacing the soil (and finding a new source of good topsoil).
- **Copper (Cu) deficiency** is characterised by small chlorotic spots at the edge of newly opened fronds. The fronds turn yellow, starting from the tip of the pinnae. The deficiency rarely appears in the nursery unless the fertiliser programme is inefficient or imbalanced. The use of peat soil as a planting medium can cause copper deficiency. Corrected by application of 0.05% $CuSO_4$ solution mixed with a general foliar fertiliser and carried out four times of application in a row.

References

Duckett, J.E. (1999) *A Guide to Oil Palm Nurseries*. The Incorporated Society of Planters, Malaysia.

Mathews, J., Tan, T.H., Yong, K.K., Chong, K.M., Ng, S.K. and Ip, W.M. (2010) Managing oil palm nursery: IOI's experience. *The Planters* 86(1016), 771–785.

Netafim (2018) *PalmDrip™ for Oil Palm Nursery Management*. Unpublished.

Rankine, I. and Fairhurst, T. (1998) *Oil Palm Nursery – Field Handbooks*. Potash and Phospate Institute, Singapore.

Rahmana, A.A., Mohebi, A.H. and Khayat, M. (2017) Study of different fertilization methods on oil palm (*elaeis guineensis*) vegetative factors. *Journal of Crop Nutrition Science* 3(1). 37 47.

Tan, C.C. (2011) *Nursery practice for production of superior oil palm planting material*. In: *Agronomic Principle and Practice of Oil Palm Cultivation*. Goh, K.J., Chiu, S.B. and Paramananthan, S. (eds), pp. 145–169. Agricultural Crop Trust, Petaling Jaya.

Tolan Tiga Indonesia (2011) *Nursery*. Tolan Tiga Indonesia Standard Operational Procedures. Unpublished.

Turner, P.D. and Gillbanks, R.A. (2003) *Oil Palm Cultivation and Management*, 2nd edn. The Incorporated Society of Planters, Malaysia.

Watering

7

Abstract

Correct watering is essential for healthy plant production. Deficient or excessive water will result in poor plant growth. Watering should avoid plant disturbance as this can lead to plant stress. This chapter describes guidelines and methods in watering application.

7.1 Watering Systems

Drip irrigation of young palm plants (seedlings or ramets) involves precise and uniform delivery of small quantities of water (and fertilisers) in each bag. This fosters fast growth and reduces water inputs and fertiliser costs. Young oil palm plants are planted and re-planted in bags filled with top-soil or a soil mixture. The young plants remain in bags until field planting. Maintaining optimal moisture, nutrients and aeration conditions in the bag throughout the pre- and main nursery stages is one of the most critical factors in producing uniform, robust and healthy plants, in the shortest time possible. Drip irrigation systems such as the PalmDrip™ Netafim technology will ensure individual care of every bag in the nursery. In each bag, a single dripper is responsible for the discharge of an accurate volume of water and (in fertigation) nutrients, at any desired interval, matching palm requirements.

7.1.1 Increased beneficial use of water

There is a general agreement that irrigation water requirements are less with drip irrigation of bags than with traditional methods – namely, fixed sprinkler systems, sprinkler hoses, rain guns, etc., where up to 90% of the water fails to enter the bags and is wasted. The daily irrigation water requirement of nursery plants under most conditions has been calculated at

Fig. 7.1. Drip irrigation for oil palm seedlings (Netafim, 2018).

6 mm/day (5.0 litres/bag/day) using traditional irrigation systems (Turner and Gillbanks, 2003). The proposed PalmDrip™ irrigation will meet the daily irrigation water needs of 11,934 palms/ha by applying water @2.4 mm/day (~2.0 litres/bag/day), a net saving of 52% water per ha, owing to the elimination of run-off and percolation losses.

- **Improved fertiliser application by fertigation.** PalmDrip™ offers considerable flexibility in fertilisation via irrigation systems. Frequent or nearly continuous application of plant nutrients in each bag along with irrigation water using Netafim FertiKit™ 3G dosing units is feasible and is beneficial for seedling and ramet growth and development. Besides fertilisers, other agro-chemicals such as herbicides, insecticides, fungicides, etc. can be supplied to improve palm growth and development (see Chapter 6).
- **Reduced operational labour.** PalmDrip™ systems can be automated easily where labour is limited or expensive. Simple automation equipment consists of electrical or battery-operated time clocks that activate pumps and solenoid valves at selected times during the day. Soil moisture sensors, computer interfacing and remote controllers can be integrated into the system.
- **Decreased energy requirements.** PalmDrip™ has the potential for reducing pumping energy costs since the operating pressures are considerably lower compared to other traditional pressurised irrigation systems. However, real energy conservation under PalmDrip™ should come from reducing the amount of water pumped because of a higher on-farm irrigation application efficiency.

- **Avoidance of soil problems.** Prevents development of substrate surface crusting and capping as observed in the case of overhead sprinkler irrigation. Water infiltration into the soil can be improved by using low-application-rate drippers.
- **Enhanced plant growth.** Slow and frequent application of water at a predetermined rate ensures fairly constant soil–water content at field capacity level in the substrate with elimination of wide fluctuations, which typically result from traditional sprinkler and hosepipe irrigation methods. Optimisation of soil water–air regime in the substrate favours uniform plant vigour, development form and leaf characteristics; improved pest and disease control; no fertiliser scorching; and less culling losses. Uniform plants help enhance production in the field 6–12 months earlier, eliminating delayed maturity and uneven palm stand.
- **Faster and higher return on investment.** Faster and uniform plant growth reduces occupancy time in the nursery, enabling the development of a field-ready plant in a shorter nursery cycle time, i.e. in 8.5 months, with more height, a well-developed canopy architecture with increased number of leaves, higher leaf area and leaf weight, larger base diameter and well-developed root system (more root mass per unit volume of soil) (Fig. 7.2). The cost of production of field-ready plants has been estimated at US$0.26 per plant considering a 10-year life of the drip system (Netafim, 2018).

Fig. 7.2. Improved leaf area and root growth in palm by drip irrigation and fertigation (Netafim, 2018).

7.1.2   Crop water requirements

The primary objective of irrigation is to provide young palm plants (seedlings and ramets) with sufficient water in bags to maintain optimum growth and development to give high-quality field-ready planting material. The required timing and amount of applied water is determined by the prevailing climatic conditions, the palm plant development stage, soil properties (such as water-holding capacity) and the extent of root development. Water within the bag substrate is the source to satisfy the plant's evapotranspiration. Thus, it is important to consider the bag water balance to determine the irrigation water requirements of nursery plants. Young palm roots require moisture and oxygen in an optimal ratio to live and develop. Where either is out of balance, owing to either over- or delayed watering, root functions are retarded and plant growth is reduced or even arrested. Therefore, to prevent any water stress, the plants should be irrigated before a given percentage of the available soil water in the root zone has been used up.

Conventionally, oil palm nursery plants are raised in bags consisting of a small volume of soil which is irrigated with fixed overhead sprinklers, rain guns or sprinkler hoses. Using sprinklers, it is difficult to irrigate each bag in the design area precisely with the same amount of water and fertiliser uniformly without either over-watering or under-watering some palms. Excess irrigation application can result in loss of applied plant nutrients through leaching and run-off from bags polluting the water resources. Inadequate irrigation application results in water stress affecting both nutrient availability and plant growth and development. Additionally, some of the water applied by sprinklers misses the bags altogether and falls on the ground surface between bags, increasing nutrient losses by run-off (this can be as much as 90%). Further, high intensity rains are expected to cause leaching of nutrients and run-off from bags. Lack of uniformity in fertilisation means more fertiliser is needed for a given number of bags in a given design area.

Estimation of crop water requirements is an important factor in irrigation scheduling to determine how much water and when to apply it. Reliable estimates of nursery oil palm water use provide a basic tool for computing water balance and predicting water availability and requirements. The term 'crop water requirement' is defined as the amount of water required to compensate the evapotranspiration loss from the cropped field. The crop evapotranspiration (ETc) refers to the combined loss of water by soil evaporation and crop transpiration from disease-free, well-fertilised palms, grown in large fields, excellently managed under optimum soil–water conditions, and achieving successful production under the given climatic conditions. Although the values for crop evapotranspiration and crop water requirement are identical, crop water requirement refers to the amount of water that needs to be supplied, while crop evapotranspiration refers to the amount of water that is lost through soil evaporation and plant transpiration, i.e. ETc. The crop water requirement is affected by a number of factors (Fig. 7.3), namely, (i) weather parameters (solar radiation, air temperature, humidity and wind speed); (ii) crop characteristics (crop type and variety, plant density, development stage, canopy cover, rooting characteristics); and (iii) management and environmental aspects (irrigation method, salinity and fertility).

The irrigation water requirement of nursery oil palm plants under most conditions has been calculated at ~6.0 mm per day under conventional irrigation systems, namely, fixed sprinkler systems, sprinkler hoses, rain guns, etc. after allowing for run-off, evaporation and evapotranspiration. Since PalmDrip™

eliminates deep percolation and run-off besides minimising soil evaporation losses (as water is precisely delivered in the bags only), the crop water requirements can be substantially reduced when irrigating with a drip system. Preliminary observations indicate that the irrigation water requirements would come down by 50%. Only @2.4 mm/day (~2.0 litres/bag/day) was estimated to be adequate under diverse agro-ecological conditions.

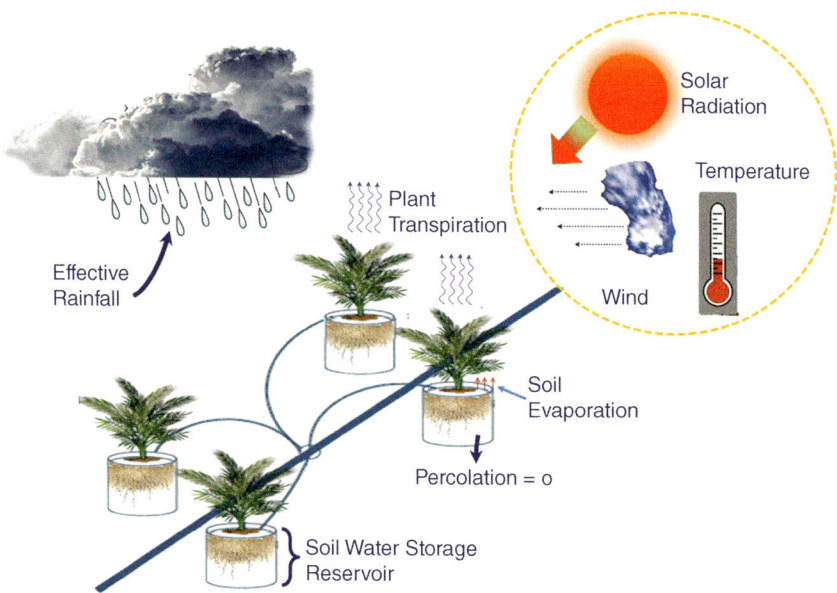

Fig. 7.3. Factors affecting water requirement of oil palm seedlings (Netafim, 2018).

7.2   Sprinkler Irrigation

7.2.1   Watering in pre-nursery

A proper irrigation system should be laid out, giving complete coverage to the whole area covered by the seedbeds. Watering can be done by using a rubber tube installed with nozzle at the head, in order to distribute the water evenly and to avoid soil erosion on the small bag. Watering in the pre-nursery is done twice a day, in the morning and afternoon, with water consumption at the rate of 0.2–0.3 l of water/day/bag.

7.2.2   Watering in main nursery

The piping system in the nursery should be properly installed (usually by a contractor) for even watering throughout the nursery. The characteristics of the water pump should be in accordance with the water requirements. Watering by sprinkler irrigation systems consumes much more water. The water should be free of fine debris that can cause blockage in the sprinkler heads, thus it is not normally suitable in peat areas due to the water containing litter and decayed wood.

Steps to calculate requirements of watering by sprinkler irrigation are described below (Tolan Tiga Indonesia, 2011).

- Water consumption per bag is 2 l/day. Calculate the water consumption for existing number of palms.
- The water per bag in sprinkler irrigation system is about 10 mm/hour. If the rainfall > 10 mm watering should not be done. If the rainfall < 10 mm or 5 mm watering should be done for only 5 mm on the day equivalent with a sprinkler pressure of 3.5 kg/cm^3.
- When the water pump machine is running at 60 m^3/h and sprinkler pressure at 3.5 kg/cm^2 it will produce 3.63 m^3 of water per hour, then the amount of sprinkler can be operated simultaneously:

$$\frac{(80\% \ x \ 60 \ m^3/h)}{3.63 \ m^3/h} = 13 \ sprinkler.$$

- If 47 sprinklers are operated, then the operation shift per day:

$$\frac{47 \ sprinkler}{13 \ sprinkler} = 3.6 \approx 4 \ times.$$

- Operation time per day (Tolan Tiga Indonesia, 2011). Nozzle at a pressure of 3.5 kg/cm^2 will produce water discharge of 6.80 mm/h. To supply the water requirements 10 mm/day per palm, the duration of operation

per rotation $= \dfrac{10 \ mm}{6.80 \ mm/hours} = 1.47 \ hours \approx 1.5 \ hours.$

Example

Plants	10,000 plants
Water requirement	2 l/seeds
Capacity/hectare	12,000 plants
Sprinkler water reach	7.5 m

Water requirement = 10,000 plants × 2 L = 20,000 L (20 m^3)

Area requirement $= \dfrac{10,000}{12,000} = 0.83 \ ha = 8,300 \ m^2$

Sprinkler coverage $= 3.14 \times 7.5 \ m \times 7.5 \ m = 176 \ m^2$

Number of sprinklers $= \dfrac{8,300 \ m^2}{176 \ m^2} = 47 \ pcs \ sprinkler$

$$\text{Water requirement per sprinkler} = \frac{20,000\,\text{L}}{47\,\text{pcs}} = 425\ \text{L/pcs}$$

Operation per day = 4 × 1.5 hours = 6 hours

Maintenance of sprinkler irrigation system

Maintenance increases the lifetime of pumps, and requires the following:

- Lubricant inside the pump should be replaced every 150 hours of operation.
- The end of suction pipes should be free from trash while pipe connections should be free from leakages. Water pumps that show signs of impairment should not be operated. If the irrigation system is temporarily deactivated, remove the standpipe and sprinkler head from the pipe. The pump should be lubricated and turned on ±30 minutes once a week.
- Sprinkler nozzle rotation should be perfect and spray radiance should be a standard distance. If these are not achieved, it is probably due to the pump pressure (working pressure 3.5 kg /cm²). Check all suction pipes as well as pipe connections to trace any leakage, or valve control distribution valve to the pipelines: which ones are open or closed?
- Check the sprinkler sprays for perfect spraying. It is important to set the duration of the operation. Install some valves in primary and secondary pipes, so that in the event of sprinkler breakdown the watering can be done manually using a rubber hose.

Fig. 7.4. Watering the oil palm seedlings using sprinkler irrigation system.

References

Gillbanks, R.A. (2003). Standard Agronomic Procedures and Practices. In *Oil Palm Management for Large and Sustainable Yields*. Fairhust, T. and Hardter, R. (Eds). Potash & Phosphate Institute, Potash & Phosphate Institute of Canada & International Potash Institute.

Netafim (2018) *PalmDrip™ for Oil Palm Nursery Management*. Unpublished.

Tolan Tiga Indonesia (2011) *Nursery*. Tolan Tiga Indonesia Standard Operational Procedures. Unpublished.

Turner, P.D. and Gillbanks, R.A. (2003) *Oil Palm Cultivation and Management*, 2nd edn. The Incorporated Society of Planters, Malaysia.

Culling

8

Abstract

Culling is a quality control procedure and as such is possibly one of the most important procedures to be carried out in the nursery. It ensures that only healthy and uniform nursery palms are maintained and these are likely to give the highest yields after planting in the field. A basic rule when inspecting plants is, 'If in doubt, pull it out!'. Off-type plants may be a result of poor growing conditions or inherent physiological or genetic disorders. Three rounds of culling are carried out during the main nursery period: the first prior to transplanting to the main nursery; the second at six months; and the third coinciding with transfer to the field or at 12 months, whichever comes earlier. This chapter describes guidelines and methods in culling application.

8.1 Considerations for Culling

Culling will reduce the number of young plants, but is essential in delivering high quality planting materials, which is the primary aim. Some plant abnormalities are hard to distinguish. In such cases the plant should be marked so that an experienced worker can make the decision to cull or not.

Bad nursery practices can lead to increased numbers of abnormal seedlings. Abnormal seedlings need to be culled as they will reduce the yield in the plantation. Basiron and Weng (2003) state that culling rates in the nursery can be up to 30%. The number of culls and culling rate can vary between plantations and is often determined by the experience of the nursery manager. Jacquemard and Zaelanie (2007) suggest that culling rates should below 15% in a well managed nursery. In general, lower than 10% is expected to result in the retention of some abnormal seedlings and culling of 25% or more reflects problems in either nursery conditions, seed quality (fungal or bacterial infection or poor handling) or the genetics of the material. Different nursery managers have reported various culling rates among populations given different result. Seed size may be a factor in subsequent seedling development and so small seed (small endosperm) may give rise to poor seedlings and higher culling rates (Murugesan, 2009).

The nursery environment aims to be homogenous, but may suffer from variation in irrigation, heterogeneous soils, depth of seed planting, drainage, experience of workers etc., thus the environmental effects will never be completely absent. Variation of performance mainly occurs due to genetic effects among and between populations. In research station nurseries, particularly in breeding trials, high culling rates will reduce undesirable seedlings caused by their genetics and can provided palms with high yield with low culling rates.

8.2   Culling in the Pre-nursery

Culling in the pre-nursery and/or rametry is the first round. This culling can be done quickly as the young seedlings or ramets are closely packed. Culling is conducted before transfer to, and potting-on, in the main nursery (three- to four-month-old seedlings and ramets).

Fig. 8.1. Experienced worker supervising the third cull and selecting high quality plants for transfer to the field.

The main categories of off-types include the following.

- **Narrow leaf.** The lamina is very narrow and the shoot can be developed like reed in extreme cases (Fig. 8.2a).
- **Crinkled leaf.** Here the leaf is corrugated transversely, with various degrees of severity (Fig. 8.2b). Crinkled leaf may be related to pest attacks or boron deficiency (Duckett, 1999).
- **Twisted leaf.** Leaf becomes coiled and twisted (Fig. 8.2c). This abnormality is often caused by incorrect planting when the plumule is planted upside down. Other causes include herbicide damage, especially from hormone-type herbicides (Rankine and Fairhurst, 1998; Duckett, 1999).
- **Rolled leaf.** The rolled leaf symptom consists of a single spike where the leaflets fail to open correctly (Fig. 8.2d).

8.3 Culling in the Main Nursery

Culling in the main nursery can be more difficult. Experience plays an important role in selection. In many cases, where there are doubts, it is normal practice to place an easily seen 1 m bamboo peg in the bag so that a more experienced worker can check the plant and take a final decision on culling.

Culling in the main nursery is carried out in two rounds. First, culling is carried out when palms are about three to four months after transfer from the pre-nursery (seedlings) or rametry (ramets), when the fronds of adjacent palms have not started to overlap. The third and final round of culling is carried out when the plants have been in the main nursery for about six to seven months.

If transfer to the field is delayed, the final round of culling must be carried out before etiolation sets in, as at this stage it is extremely difficult to distinguish between abnormal and normal palms.

Typical characteristics of abnormal plants in the main nursery are as follows:

- **Runts.** These plants resemble normal palms but are small compared to others in their group of similar age (Fig. 8.2e). Runts are more susceptible to disease such as *Curvularia* spp. (Rankine and Fairhurst, 1998).
- **Upright.** These abnormal palms have a definite upright appearance with rigid looking fronds. The frond angle is abnormal and often the frond bases are spaced from each other (8.2f). Such plants are often sterile and thus unproductive (Rankine and Fairhurst, 1998).
- **Flat top.** Newly emerging fronds are seldom longer than the old fronds, giving the appearance of a flat top (Fig. 8.2g). Potassium (K) application may reduce this abnormality (Rankine and Fairhurst, 1998).
- **Juvenile habit.** Older fronds of main nursery plants should be divided after a few months. However, in this abnormality the leaf remains undivided (Fig. 8.2h). Such juvenile plants may be late in reaching maturity

and may be sterile (Rankine and Fairhurst, 1998), but can be useful as
ornamentals.

- **Limp form.** Normally fronds are held erect. In this abnormality the
  fronds are limp and hang-down (Fig. 8.2i). These plants are often short
  but can be useful as ornamentals.
- **Short internodes.** In palms with short internodes, the pinnae look com-
  pressed, giving a crowded appearance (Fig. 8.2j). This is thought to be a
  genetic disorder (Rankine and Fairhurst, 1998).
- **Wide internodes.** Here the pinnae are spaced wide apart with a resultant
  wide-open canopy (Fig. 8.2k). This may occur due to etiolation in nar-
  row-spaced nurseries. However, in normal-spaced nursery plants this
  symptom is likely to be a genetic abnormality (Rankine and Fairhurst,
  1998; Duckett, 1999).
- **Narrow pinnae.** This abnormality is marked by very narrow pinnae,
  which form a sharp point. Plants often have a pale green appearance

Fig. 8.2. a) Narrow leaf; b) Crinkled leaf; c) Twisted leaf; d) Rolled leaf;
e) Runts; f) Upright or sterile; g) Flat top seedings; h) Juvenile seedlings; i) Limp
form; j) Short internodes; k) Wide internodes; l) Narrow pinnae; m) Chimera;
n) Collante.

(Fig. 8.2l). Narrow pinnae is a genetic disorder (Rankine and Fairhurst, 1998).

- **Chimera.** Usually due to genetic factors (Duckett, 1999). Palms show chlorotic or white leaf stripes (Fig. 8.2m). Symptoms normally appear after four months after potting-on in the main nursery. An abnormal incidence should be reported to the supplier and plants should be replaced.
- **Collante.** Plants display marked constriction around the central part of the lamina, which prevents full leaf expansion (Fig. 8.2n). Induced collante symptoms may occur within a few weeks after germination or transplanting, sometimes after a period of normal growth. Pathological or environmental factors (especially moisture stress) which impair root development may cause collante symptoms (Duckett, 1999). This disorder is reversible with careful and regular watering (Rankine and Fairhurst, 1998) and culling should be delayed until plants are about six months old, by which time induced collante symptoms may have disappeared.

8.4 Culling Ramets

In addition to the off-type symptoms described above, ramets may display somaclonal variation, due to tissue culture. The most common type of abnormality is the 'truncated leaf' or 'self-pruning' leaf symptom. Severely affected ramets in the rametry will be stunted in growth and may die.

Poor management and agronomic practices, particularly inadequate watering and lack of humidity, can cause an apparent increase in seemingly defective palms, e.g. grass leaf palms, which can lead to unnecessary losses. Ramets should be supplied from a reputable producer and have a strong root system.

References

Basiron, Y. and Weng, C.W. (2003) Going Back to Basics: Producing High Palm Oil Yields Sustainably. *Oil Palm Bulletin* 46(May 2003): 1–14

Duckett, J.E. (1999) *A Guide to Oil Palm Nurseries*. The Incorporated Society of Planters, Malaysia.

Jacquemard, J.Ch. and Zaelani, H. (2006) How to choose the best planting material to obtain a maximal crop. Paper presented to PT Socfindo Tenera Gathering and Tour, 2007. Medan, Indonesia. 11–13 June 2007. Available at: https://agritrop.cirad.fr/554884/1/document_554884.pdf (acessed 5 May 2019).

Murugesan, P. (2009) Effect of seed size on germination and seedling growth of oil palm hybrids. Proceedings of Agriculture, Biotechnology & Sustainability Conference, PIPOC 2009. Volume 2: Poster Papers.

Rankine, I. and Fairhurst, T. (1998) *Oil Palm Nursery – Field Handbooks*. Potash and Phosphate Institute, Singapore.

Weeding

9

Abstract

Various weed species are common in the main nursery and include grasses and sedges as well as broad-leafed plants. The main nursery is susceptible to weeds, especially wind-dispersed species, as there are open spaces between the bags. Weeds are managed using mechanical and chemical methods. It is important to keep the bags free from weeds, as these will compete for moisture, nutrients and sunlight. Weeding also provides hygienic conditions in the nursery that help avoid pest or disease problems. Weeding is carried out within the bags and in the inter-row spaces. Weeding is included in standard nursery management activities. This chapter describes guidelines and methods for weeding.

9.1 Materials and Tools

Materials for weeding activities:

- Knapsack sprayer with fitted nozzle shield and labelled with the word 'herbicide'.
- Hoe.
- Wheelbarrow to carry away the weeds.
- Measuring cup and stirrer.
- Personal protective equipment (PPE).
- Herbicide.
- Adhesive material or surfactant.
- Water.

9.2 Chemical Weeding

Herbicide should be used in the main nursery for a double-stage nursery and after four months for single-stage nursery. Chemical weeding of the bag is not recommended. Direct control of weeds by herbicide spraying may be done, but care must be taken to ensure safety of the operator (see Chapter 2) and the plants. Attention should be paid to the following:

- Correct type of herbicide, which does not damage plants by root uptake. A contact herbicide is recommended.
- Avoid hormone-based herbicides in and around the nursery as these may have serious effects on palms. Water catchment areas serving the nursery should also be free of hormone-based herbicides.
- A garbage area should be placed nearby, and materials incinerated once a week to prevent pest and disease spread.
- Leaves that have been in contact with the herbicide may show orange spots or dead tissue two days after herbicide application. The palms can recover when the herbicide used is a contact herbicide. Plants cannot recover from damage by systemic herbicides (e.g. glyphosate).
- The nursery manager should routinely check and take appropriate fast action when workers act carelessly.

Methods for chemical weeding are described below.

- Fill the knapsack sprayer with 15 l of water.
- Install the nozzle of the sprayer correctly to set the coverage area of herbicide. A shield should be fitted to the nozzle head. Take care when spraying to avoid spray-drift to palms (be aware of wind).
- The sprayer should be set to cross the row.
- Measure out the herbicide (measuring cup) according to the recommended dosage (usually found on the packaging).
- Pour the herbicide into the knapsack sprayer and mix with a stirrer.
- Spray the target.
- Spraying activities need to be advertised. A warning board should give information on spray treatments, including the area and timing so that casual ingression during this time is avoided.

9.3   Manual Weeding

Methods of weeding inside the bag vary depending on bag size. Weeding can be done manually using a hoe or a rake.

- **Small bag.** Any weeds present inside the bag can easily be removed by hand. It is generally recommended that all herbicides should be avoided at the pre-nursery and rametry stage as young plants can easily be damaged and mistakes can result in large losses of plants.
- **Large bag.** Mulching can be performed to control weeds. Mulching with palm kernel shells can reduce the number of weeds and weed growth (Duckett, 1999).

Fig. 9.1. a) Weed inter-row by hoeing; b) Inter-row clear of weeds; c) Spraying workers wear complete PPE and a warning board should be placed beside the work area; d) Difference between weeding using herbicide spraying (left side of picture) and manual weeding by hoe (middle).

Reference

Duckett, J.E. (1999) *A Guide to Oil Palm Nurseries*. The Incorporated Society of Planters, Malaysia.

Pests and Diseases

10

Abstract

Pests and diseases should be avoided as they have the potential to cause high economic losses in oil palm nurseries. It is therefore important to recognise pests and diseases as soon as possible and take action. Routine and regular nursery inspections by experienced staff are essential. This chapter describes the current major pests and diseases in the oil palm nursery and their management.

10.1 Common Pests in the Nursery

- Crickets (*Brachytrupes, Acheta, Gryllus* spp.). Crickets have been pests of nursery oil palm plants for many years (Hill, 1983). They frequently attack the shoot-tip immediately after germination when the young shoot begins to emerge. Mole crickets also attack the roots and cause severe damage and rapid death. The presence of this pest is marked by ridges on the soil surface made by burrowing. Pre-nursery seedlings and ramets are particularly susceptible to crickets as the bags are close-packed (Duckett, 1999).
- Cockchafers (*Apogonia* and *Adoretus compressus*). Two species that commonly occur are *Apogonia* and *Adoretus compressus* (Hill, 1983; Duckett, 1999; Turner and Gillbanks, 2003). Invasion is usually confined to the edges of the nursery, but in severe cases the whole nursery can be affected. The beetles are nocturnal (they fly and feed at night and rest in the soil during the heat of the day). Attack by cockchafers usually occurs when the palms are four to five months old or more. Severe damage can occur to leaf tissue under certain conditions. *Apogonia* eats the marginal tissues, whereas the *Adoretus* beetle eats the leaf centre, but they do not attack the veins (thus leaving uneaten vascular tissues).

- Aphids (*Cerataphis, Hysteroneura* spp.). Invasions of aphids in the nursery are not normally significant. However, population explosions can occur at any time (Duckett, 1999). Aphids are normally found in the newly emerging frond axils. Aphid attacks on leaf tissue can cause leaf and spear distortion.
- Red spider mite (*Tetranychus piercei and Oligonychus* spp.). Red spider mite is commonly found in the oil palm nursery and is a significant pest (Duckett, 1999; Turner and Gillbanks, 2003; Mathews *et al.*, 2010). Although very small, the mites can be seen with the naked eye on the underside of the leaf. A magnifying glass can be used to see the eggs (as groups of small red dots) and the larvae. Leaf-piercing and sap-sucking from red spider mite can cause severe damage, especially during periods of dry weather. The application of pesticides may kill predators allowing this pest to increase rapidly (Wood, 1971; Turner and Gillbank, 2003). Heavy invasion will lead to discolouration and yellowing of the leaf. Symptoms of attack are similar to magnesium deficiency and inspection for the pest should be made. Red spider mites are found in a wide range of plants, particularly rubber. In mature oil palm, the mites are frequently found in pollen.
- Mealy bugs (*Dysmicoccus brevipes*). Mealy bugs are not a common nursery pest but are occasionally found in quantity on individual palms with no significant economic effect (Duckett, 1999). The bugs are waxy-white in appearance and suck the sap from the leaf tissue, causing general weakening of the palm. They are often attended by ants. Although the threat is low, severe root infestations occur periodically, with symptoms similar to nitrogen deficiency (Turner and Gillbanks, 2003; see also Chapter 6). Thus root attack should be examined in several palms.
- Caterpillars. The common species that cause damage in oil palm nurseries is *Spodoptera litura* (Turner and Gillbanks, 2003; Corley and Tinker, 2015). These caterpillars are nocturnal and eat the foliage of young plants, leaving it full of holes.
- Ants. Ant attacks in nurseries are infrequent but they can cause severe economic loss. Ants attack the plumule when the seeds are newly-planted. Losses caused by ants can be up to 20% (Turner and Gillbanks, 2003).
- Grasshoppers (*Valanga sp.*). These commonly attack plants in the nursery, especially during dry periods (Wood, 1971), but attacks are normally of little importance. However, grasshoppers can cause significant losses when they attack leaf tissues from the edges inwards. Small palm stems are occasionally attacked close to the ground.
- Weevils (*Temnoschoita* spp.). Weevils frequently attack nursery plants in West Africa (Turner and Gillbanks, 2003). *Hypomeces squamosus* more frequently attack in Sumatra, Indonesia. The weevils lay their eggs on damaged petioles. Once hatched the larvae tunnel through the tissues of the apical bud. Palms are killed when the damage is severe.

Fig. 10.1. Aphid invasion.

Fig. 10.2. Caterpillar invasion.

Fig. 10.3. Grasshopper on the oil palm seedling leaf.

- **Rats.** Attacks by rats can be frequent, especially following transfer to the field. Losses in the nursery can be severe, especially of older palms. Rats kill young palms rapidly by gnawing the palm bulb to reach the bud tissues. Rats can be a problem when the nursery is close to areas with high rat populations such as housing or mature oil palm fields (Turner and Gillbanks, 2003).

10.2   Common Diseases in the Nursery

- **Blast (*Pythium splendens*).** Blast is the most serious infection, causing high economic loss in oil palm nurseries (Duckett, 1999; Turner and Gillbanks, 2003; Corley and Tinker, 2015). Recovery from infection takes a long time. Infections are normally associated with an inefficient irrigation system (the fungal invasion is assumed to occur when the soil in the bag becomes hot and dry, Duckett, 1999) Young plants are more susceptible to infection when they are six to eight months old (Turner and Gillbanks, 2003). Symptoms can be observed in both roots and leaves. The first symptoms normally show in the oldest leaves (occasionally spear rot also occurs). Leaves change in colour rapidly from olive-green to a clear yellow, then produce purplish necrotic patches. Leaves gradually become brittle and brown. In the roots, the tissues between the central vascular system and the hypodermis are destroyed.
- ***Curvularia* seedling blight (*Curvularia eragrostidis*).** Infections of *Curvularia* usually occur during young plant establishment in the main nursery. The symptoms of the disease can vary, depending on the severity of the infection. Normally, small, dark brown spots or lesions with a distinct yellow to orange colour appear over the leaf surface. In severe cases, large lesions can appear, causing the whole leaf or leaves to die back (Duckett, 1999).

Fig. 10.4. a) *Curvularia* symptoms; b) severe infections of *Curvularia*.

- ***Corticium.*** This disease is spread through contact between infected and healthy leaves, with most infections resulting from splashing during heavy rain that carries soil particles containing the fungus to the leaves. Poor drainage is therefore a factor (Turner and Gillbanks, 2003). Infections occur most severely in the youngest leaves, where the spear leaf is infected (Turner and Gillbanks, 2003). Corticum can be a serious disease in the main nursery (Duckett, 1999). The symptoms include rows of dark brown lesions, which dry out, leaving the tissue a greyish to white colour, with a purplish brown margin.
- ***Helminthosporium.*** This disease is often associated with poor (insufficient) spacing of bags which restricts air movement between the palms (Duckett, 1999). The symptoms are dark brown spots which gradually turn yellowish, followed by die-back of the leaf, beginning at the leaf margins. The lesions then enlarge, coalesce and kill the leaf.
- **Spear or bud rot.** Infected palms show yellowing of new spears and recently opened fronds. The spear is easily pulled out and has a bad odour. The basal tissue of the frond becomes rotten and turns soft.
- **Early leaf disease.** The source of the inoculum is assumed to be palm fronds used for shading. The infection is pronounced in nursery beds under heavy shading and with high humidity (Duckett, 1999; Rankine and Fairhurst, 1998). Early symptoms are the appearance of many small, pale spots on the leaves. These spots turn brown and are followed by the death of the leaves. The spots are clearly defined with distinct borders between the healthy and infected tissues.

10.3 Pest and Disease Management Control

Pest and disease control measures often involve the use of chemical agents. Spraying treatments are a routine activity in oil palm nurseries. Effective spraying relies on accurate determination of the scale of the infestation. The treatment of spray rounds can be reduced if there is no evidence of infestation/infection. The nurseryman must be vigilant and take immediate action in the event of either minor or major outbreaks of pests or diseases (Rankine and Fairhurst, 1998).

Preventative husbandry is needed to avoid some problems, especially pests and diseases resistant to chemical agents. Clipping and removal of the damaged tissues is recommended (Duckett, 1999).

Identification of pests and diseases needs an experienced observer and the nursery worker needs to be trained to recognise them, otherwise a pathologist should be consulted (Mathews *et al.*, 2010). Training in pest and disease recognition should be arranged routinely, with refresher courses once a year. Such training provides an 'early warning' system on any outbreak by reports from people working full-time in the nursery and is the best system for monitoring plant health (Duckett, 1999). Training should also be given on the use of pesticides, fungicides, etc. As with any chemical usage, great attention should be paid to all recommendations regarding the health and safety precautions for the user (see Chapter 2) and any recommendations given by the supplier. Fig. 10.5 shows safety clothing including goggles, mask, safety boots, gloves, hat and apron, which must all be worn when spraying with fungicides and pesticides.

Fig. 10.5. Sprayer operator wearing complete PPE.

Other precautions and actions to control pests and disease in the pre-nursery are given by Mathews *et al.* (2010) and include the following:

- Maintaining good vigour and health of young plants by providing optimal water and nutrients. Vigorous plants are more resistant to disease infection.
- Ensuring absence of stagnant water in nursery beds to prevent blast disease caused by *Pythium*.
- Ensuring soil for the growth medium is taken from a place that is free from *Ganoderma* incidence.
- Avoiding mulching nursery ground with oil palm mill by-products, e.g. shells, which can promote disease infestation.
- Immediate isolation of infected plants for special chemical treatment. Quickly potting-on and transplanting pre-nursery seedlings or ramets infected by *Curvularia eragrostidis* to large bags in the main nursery was found to be very effective in controlling the disease.

10.3.1   Control of pests

- **Crickets.** Control is normally done by scattering insecticide on the soil or spraying the inter-bag spaces with the active ingredient that passes into the soil through normal spraying. Indirect control can be done by controlling the ant population because crickets eat ants (Rankine and Fairhurst, 1998).
- **Cockchafers** are nocturnal, thus insecticide application should be conducted during the late afternoon to avoid insecticide evaporation (Rankine and Fairhurst, 1998; Duckett, 1999). Preferably, the insecticide should be systemic with the residual effect occurring up to 10 days after application. Repeat applications at 10-day intervals should be made until there is no more damage. An application of granular soil insecticide may also be made, broadcasting over the ground surface of the nursery and the soil surface of the bags.
- **Aphids** can be controlled directly or indirectly. An insecticide spray directed down into the palm spear and leaf axils will quickly deal with aphids. Indirect control can be done by eliminating ants because ants protect aphids (Turner and Gillbanks, 2003).
- **Red spider mite** is an arachnid, thus controls use arachnicides to kill larvae or eggs (the adults are difficult to control with chemicals). Arachnicides should be sprayed under the leaf, ensuring the undersides of all leaflets are wetted by the spray. Red spider mite can spread rapidly, thus careful observation for the pest and early recognition and treatment are necessary (Duckett, 1999; Turner and Gillbanks, 2003).
- **Mealybugs** can be controlled by insecticidal spray or by sprinkling granules or powder of persistent insecticides around infected palms.

Alternatively, the soil may be drenched with an insecticide solution. Indirect control can be done by eliminating ants because mealybugs eat ants (Turner and Gillbanks, 2003; Verheye, 2010).

- **Caterpillars** may be controlled by pesticide spraying. When the infestation is light, they may also be controlled simply by hand-picking (Turner and Gillbanks, 2003).
- **Ants** may be controlled by insecticides with long-term persistence (dust or granular). Controlling the ant population has secondary positive effects on controlling crickets, mealybugs and aphids as described above.
- **Grasshoppers** are controlled by insecticide application at 10- to 14-day intervals until the pest is under control. Baits (that control crickets) may also be used. These consist of a mixture of bran and sawdust and an insecticide (Turner and Gillbanks, 2003).
- **Weevils** are usually controlled using preventative methods. Good nursery techniques such as maintaining hygiene (with removal and destruction of any attacked plants) can control weevils (Turner and Gillbanks, 2003).
- **Rats** are commonly controlled by bait methods. Rats are neophobic and follow the same pathways. Baiting is normally conducted along rat paths, e.g. the border across the invasion area, using anticoagulant baits (Turner and Gillbanks, 2003). Daily inspection of bait traps is needed with rat attacks as they cause rapid and devastating damage to young plants. Replace with new bait as required. Barn owls can help keep rat populations under control and nest boxes can be set up to encourage these birds (Duckett, 1999).

10.3.2 Control of diseases

- **Blast** may be controlled in various ways. Keep the soil medium moist by ensuring proper irrigation and soil–water retention (use of a good soil mixture) (Duckett, 1999; Turner and Gillbanks; 2003; Corley and Tinker, 2015). Mulch to keep the soil cool. This reduces the risk of blast (Rankine and Fairhurst, 1998; Duckett, 1999; Turner and Gillbanks 2003). Infected plants should be uprooted and incinerated as they seldom recover (Duckett, 1999; Turner and Gillbanks, 2003). The temperature of soil must be kept as low as possible by aligning the bags in east–west rows. Good nursery husbandry in keeping weeds under control will help (Duckett, 1999; Turner and Gillbanks, 2003). Understanding the planting time with the occurrence of blast disease to ensure that the blast season has passed before the palms reach the susceptible stage (Aderungboye and Esuruoso, 1976; Corley and Tinker, 2015).
- *Curvularia* is usually controlled by routine fungal spraying, normally conducted twice a month. If infections become severe, the spraying should be increased (Turner and Gillbanks, 2003). It is important to

remove all suspect plants, such as runts and those of poor vigour as these are normally the first to be attacked. Thus routine culling should be practised. Infected plants, dead and dying tissues should be removed immediately and incinerated (Rankine and Fairhurst, 1998). Plants in the nursery should be sprayed to run-off point, with particular care taken to obtain good coverage of the underside of the leaf. Routine inspection is the best way to monitor disease spread.

- *Corticium* may be controlled directly or indirectly. Routine fungal spraying is conducted every 10–14 days. Indirect control is achieved by good management of the watering system as over-watering and excessive shade should be avoided (Turner and Gillbanks, 2003). The disease is spread by water splash and excess water, and standing water should be eliminated. Remove and incinerate infected leaves as this will reduce the risk of disease spread (Rankine and Fairhurst, 1998).
- *Helminthosporium* is controlled by routine fungal spraying. The disease may be prevented by maintaining good inter-bag spacing in the nursery (Rankine and Fairhurst, 1998).
- **Spear or bud rot** control can be achieved by drenching the spear with a fungicide solution. In practice, if no new spear emerges within three to four weeks the plant should be removed and incinerated (Duckett, 1999).
- **Early leaf disease** is controlled by routine fungal spraying. Infected leaf tissues should be removed using scissors and incinerated (Duckett, 1999). Shade materials such as palm fronds are often a source of early leaf disease and should be sprayed with both insecticide and fungicide before use (Turner and Gillbanks, 2003). Reducing the shade and maintaining the air flow can control the disease (Rankine and Fairhurst, 1998).

References

Aderungboye, F.O. and Esuruoso, O.F. (1976) Ecological Studies on *Phytium splendens* Braun in Oil Palm (*Elaeis Guineensis* Jacq.) plantation soils. *Plant and Soil* 44, 397–406.

Corley, R.H.V. and Tinker, P.B. (2015) *The Oil Palm*, 5th edn. Wiley Blackwell, UK, p. 639, ISBN: 978-1-405-18939-2.

Duckett, J.E. (1999) *A Guide to Oil Palm Nurseries*. The Incorporated Society of Planters, Malaysia.

Hill, D.S. (1983) *Agriculture Insect Pest of the Tropics and Their Control*. Cambridge University Press, UK, p.749, ISBN: 0-521-28867-3.

Mathews, J., Tan, T.H., Yong, K.K., Chong, K.M., Ng, S.K. and Ip, W.M. (2010) Managing oil palm nursery: IOI's experience. *The Planters* 86(1016), 771–785.

Rankine, I. and Fairhurst, T. (1998) *Oil Palm Nursery – Field Handbooks*. Potash and Phospate Institute, Singapore.

Turner, P.D. and Gillbanks, R.A. (2003) *Oil Palm Cultivation and Management*, 2nd edn. The Incorporated Society of Planters, Malaysia.

Verheye, W. (2010). Growth and production of oil palm. In: Verheye, W. (ed.), *Land Use, Land Cover and Soil Sciences. Encyclopaedia of Life Support Systems (EOLSS).* UNESCO-EOLSS Publishers, Oxford, UK. Available at http://www.eolss.net (accessed 22 March 2019).

Wood, B.J. (1971) Development of integrated control programs for pests of tropical perennial crops in Malaysia. In: Huffaker, C.B. (ed.) *Biological Control.* Plenum Press, New York, USA.

Pre-field Genotypic Screening and Selection

11

Abstract

The advent of DNA screening in oil palm allows plants in the nursery to be screened for specific traits prior to potting-on and field planting, thus saving space and resources in the nursery and maximising field productivity by eliminating undesired genotypes. This has become particularly important in screening for shell type, i.e. in distinguishing Dura, Pisifera and Tenera palms at an early stage. In this chapter, we describe some of the basic features in DNA analysis from leaf sampling in the nursery to DNA extraction in the lab and various DNA analyses that may be performed either in-house or outsourced. The oil palm genome was sequenced in 2013 (Singh *et al.*, 2013) and this has paved the way to develop DNA diagnostics for traits of interest. This is a fast-developing area. The chapter represents an introduction to DNA markers in oil palm and their application in the nursery. It does not provide step-by-step protocols, as this is the subject of another manual in this series (Anwar *et al.*, in preparation).

11.1 Traits to be Scored

A major breakthrough in oil palm genetics came in 2013 when scientists from the MPOB announced and published the genome sequence of the oil palm (Singh *et al.*, 2013). This provided new opportunities in genetics (e.g. understanding gene function), plant breeding (e.g. marker-assisted selection) and commercial production (e.g. quality control).

There are several traits of particular current interest in oil palm: shell thickness (a major determinant of oil yield), *virescens* (fruit colour), mantled fruit (off-type flowering), *Ganoderma* resistance (a major disease in Southeast Asia), *Fusarium* resistance (a major disease in Africa), drought tolerance (adaptation to climate change and drier environments), long stalk (for ease in harvesting), slow height increment (allows plantations to be productive for longer), oil quality (potential new markets) and nutrient-use efficiency (reduces costs in fertiliser application).

The shell thickness gene (*Sh*) is often considered the most important gene in oil palm production. It controls the fruit type: the Dura type has a thick-shelled kernel (*Sh/Sh*), Pisifera has no shell (*sh/sh*) and Tenera is thin-shelled (*Sh/sh*), this is important as the commercial type is Tenera which is a product of crossing Dura with Pisifera (see Setiawati *et al.*, 2018 and Kelanaputra *et al.*, 2018 for manuals on crossing and seed production in oil palm). The mapping and identification of the *Sh* gene was reported by Singh *et al.* (2013). Without genetic markers, shell type in oil palm can only be determined once fruit is produced (five years from sowing). This has been a great hindrance in oil palm development as, for example, in a cross between Tenera x Tenera only 50% of the progeny will be Tenera, 25% will be Dura and 25% will be Pisifera (which is normally sterile). Screening in the nursery for *Sh*-determination now allows greater efficiency in selecting specific types for breeding trials (Sitepu *et al.*, 2019).

Fruit colour is another trait that can be readily screened for by geno-typing. The *virescens (Vir)* gene controls fruit skin (exocarp) colour and is an indicator of ripeness. Breeders are therefore interested in *Vir*, and like *Sh* the phenotype is only visible once palms produce fruit. Thus a DNA diag-nostic for *Vir* allows plants to be selected in the nursery.

Mantling of fruit is another trait of interest to breeders and planters. This is an epi-genetic trait produced mainly in tissue culture. Mantled fruits are degenerate and do not yield oil, thus it is important to screen out all palms carrying this defect before field planting. A simple genetic test such as that developed by MPOB and Orion named SureSawit™ KARMA allows palms to be screened for the mantled disorder (Ong-Abdullah, *et al.*, 2016), and this may be applied to plants in the nursery (or earlier while they are still in culture).

Other traits will soon be developed for marker-assisted selection, but in addition to screening positively for specific traits, breeders also aim to main-tain the genetic background of successful commercial varieties. Thus screen-ing for the genetic background (genetic fingerprinting) is also of interest. DNA testing can also be used in quality control and legitimacy testing.

11.2   DNA Extraction

DNA is composed of four nucleotides: adenine (A) and guanine (G) are the larger purines; cytosine (C) and thymine (T) are the smaller pyrimidines. Evaluation of genomic DNA sequences is now fundamental for both func-tional genomics and marker-assisted breeding. The starting point for this is the extraction of high-quality DNA. The extraction of genomic DNA requires careful sample preparation, followed by tissue lysis and isolation of the DNA (Fig. 11.1).

Fig. 11.1. Workflow for DNA extraction: leaf sample collection from the nursery, leaf preparation, tissue maceration and DNA extraction.

Ideal DNA extraction methods should be high-throughput, low cost, high DNA yield per mg of tissue and limited/zero use of toxic chemicals. Commercial kits have been developed for DNA extraction. These are suitable for rapid and reliable DNA extraction of high purity and high quality, and produce high yields.

Following DNA extraction, it is necessary to assess the quality and quantity of DNA. There are many methods, such as gel-based electrophoresis, spectroscopy and fluorimetry. For quantitating DNA or RNA with a spectrophotometer, readings should be taken at wavelengths of 260 nm and 280 nm. The reading at 260 nm allows calculation of the concentration of nucleic acid in the sample. The reading at 280 nm gives the amount of protein in the sample.

    1 O.D. at 260 nm for double-stranded DNA = 50 ng/ul of dsDNA
    1 O.D. at 260 nm for single-stranded DNA = 20-33 ng/ul of ssDNA
    1 O.D. at 260 nm for RNA molecules = 40 ng/ul of RNA

Pure preparations of DNA and RNA have OD260/OD280 values of 1.8 to 2.0, respectively. If there is contamination with protein or phenol, this ratio will be significantly less than the values given above, and accurate quantitation of the amount of nucleic acid will not be possible (Fig. 11.2).

Plant breeders require genotypic analyses of large numbers of individuals for diversity analysis, marker-assisted selection, variety fingerprinting and legitimacy testing. Thus, DNA extraction must be of sufficient quantity and quality to generate robust and easily scored data for a number of analyses.

11.3   DNA Testing

The range of downstream applications for DNA continues to grow with technology advancement. Many applications are now PCR based (Roux, 2009), and multiplexing to identify more than one target per reaction is frequently performed. For many applications, a fluorescent label is incorporated during the PCR or in a separate reaction to aid in detection and

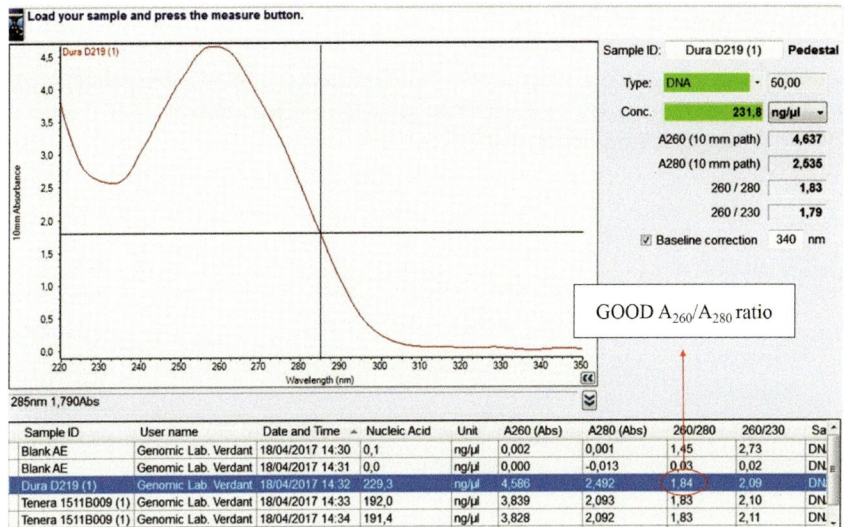

Fig. 11.2. Assessing the quality and quantity of DNA using spectrophotometer.

analysis of specific DNA sequences. Many of these applications are based on real-time or quantitative PCR (Wittwer *et al.*, 2012). Increasingly, sequencing applications are based on fluorescence and analysed following capillary electrophoresis or hybridisation (Sultana and Khan, 2007; Durney *et al.*, 2015).

As functional genomics increases in popularity, so does the desire to isolate multiple molecules (e.g., genomic DNA, RNA and protein) from a single sample and correlate the presence and levels of gene products with specific gene alleles. DNA markers and DNA diagnostics allow for greater efficiency in selection in plant breeding by nursery screening.

In addition to the in-house protocols, such as that described above, *Sh* determinations can be done using outside services such as Orion Biosains (www.orionbiosains.com). Orion Biosains also provides services to screen for virescens fruit (*Vir*, Singh *et al.*, 2014).

11.4   Future Pre-field Screening

DNA approaches and tools have great potential to optimize plant breeding efforts. In recent years, there has been an exponential increase in DNA resources and methods aimed at identifying DNA variation (polymorphism) which controls various traits of interest.

Today, significant efforts are being made in several countries to develop new DNA resources (e.g. sequencing, re-sequencing and chip development; Pareek *et al.*, 2011) and there are several advancements in high-throughput genotyping. Progress in genotyping has taken a giant leap forward with the increasing ease of DNA sequencing (Shu *et al.*, 2012). A range of 'next generation' sequencing platforms exist, e.g. sequencing by synthesis, ion semi-conductor, single-molecule real-time sequencing, nanopore sequencing and other sequencing technologies (Buermans and den Dunnen, 2014; Rival, 2017). Thus, the stage is set for ever-increasing genotypic selection in the nursery, and nursery practices will inevitably change to accommodate this.

References

Anwar, Y. *et al.* (in preparation) *DNA Markers in Oil Palm: A Manual. Techniques in Plantation Science.* Forster, B.P. and Caligari, P.D.S. (eds). CAB International, Wallingford, UK, in press.

Buermans, H.P.J. and den Dunnen, J.T (2014) Next generation sequencing technology: Advances and applications. *Biochimica et Biophysica Acta* 1842(10), 1932–1941, DOI: 10.1016/j.bbadis.2014.06.015.

Durney, B.C., Crihfield, C.L. and Holland, L.A. (2015) Capillary electrophoresis applied to DNA: determining and harnessing sequence and structure to advance bioanalyses (2009–2014). *Anal Bioanal Chem* 407, 6923–6938, DOI 10.1007/s00216-015-8703-5.

Kelanaputra, E.S., Nelson, S.P.C., Setiawati, U., Sitepu, B., Nur, F., Forster, B.P. and Purba, A.R. (2018) *Seed Production in Oil Palm: A Manual. Techniques in Plantation Science.* Forster, B.P. and Caligari, P.D.S. (eds). CAB International, Wallingford, UK, p. 59.

Ong-Abdullah, M., Ordway, J.M., Jiang, N., Ooi, S.E. and Martienssen, R.A. (2016) *SureSawit KARMA-A Diagnostic Assay for Clonal Conformity.* MPOB TS No. 156.

Pareek, C.S., Smoczynski, R. and Tretyn, A. (2011) Sequencing technologies and genome sequencing. *Journal of Applied Genetics* 52, 413–435, DOI 10.1007/s13353-011-0057-x.

Rival, A. (2017) Breeding the oil palm (*Elaeis guineensis* Jacq.) for climate change. *OCL*, 24(1): D107.

Roux, K.H. (2009) *Optimization and Troubleshooting in PCR*, 4(4). Cold Spring Harbor Laboratory Press, DOI:10.1101/pdb.ip66.

Setiawati, U., Sitepu, B., Nur, F., Forster, B.P. and Dery, S. (2018). *Crossing in Oil Palm: A Manual. Techniques in Plantation Science.* Forster, B.P. and Caligari, P.D.S. (eds). CAB International, Wallingford, UK, pp 96.

Shu, Q.Y., Forster, B.F. and Nakagawa, H. (2012) *Plant Mutation Breeding and Biotechnology.* CAB International, Wallingford, UK, p. 608.

Singh, R., Ong-Abdullah, M., Low, E.T.L., Manaf, M.A.A., Rosli, R., Nookiah, R., Sambanthamurthi, R. *et al.* (2013) Oil palm genome sequence reveals divergence of interfertile species in old and new worlds. *Nature*, DOI: 10.1038/nature12309.

Singh, R., Low, E.T.L., Ooi, L.C.L., Ong-Abdullah, M., Nookiah, R., Sambanthamurthi, R. *et al.* (2014) The oil palm VIRESCENS gene controls fruit colour and encodes a R2R3-MYB. *Nature Communications*, DOI: 10.1038/ncomms5106

Sitepu, B., Setiawati, U., Nur, F., Laksono, N.D., Anwar, Y. *et al.* (2019) *Field Trials in Oil Palm Breeding: A Manual. Techniques in Plantation Science.* Forster, B.P. and Caligari, P.D.S. (eds). CAB International, Wallingford, UK, in press.

Sultana, G.N.N. and Khan, A.H. (2007) Optimization of the sample preparation method for DNA sequencing. *Journal of Biological Sciences*, 7, 194–199.

Wittwer, C., Hahn, M. and Kaul, K. (2012) *Rapid Cycle Real-Time PCR – Methods and Applications: Quantification.* Springer, DOI: 10.1007/978-3-642-18840-4.

Quarantine Nurseries **12**

Abstract

Although Indonesia is the largest grower of oil palm, other countries are also significant producers. Malaysia, Papua New Guinea, Costa Rica and Thailand are the next biggest oil palm seed producers. Due to this competition, some Indonesian plantation companies import oil palm germinated seeds from foreign producers. In addition, in order to widen the oil palm germplasm, especially for breeding purposes, countries engage in exchange or collection consortia and this can involve not only different countries, but also different continents, e.g. Africa, South America and Southeast Asia. One consequence of such import programmes is that each recipient of foreign germplasm must adhere to quarantine regulations, which are set up to protect the local species and commercial plantations by preventing the introduction of foreign pests and diseases. The regulations are usually issued and monitored by national governmental bodies, such as the Ministry of Agriculture, with inspections conducted by official quarantine agencies. Rules, regulations and processes vary from country to country. Here we give those that apply to Indonesia.

12.1 Pests and Diseases in the Nursery

The Crop Protection Compendium (CPC) provides a comprehensive list of crop pests and diseases in the world (https://www.cabi.org/cpc), it provides data sheets on pests and diseases, their mode of transfer and geographic distribution. Also each country has its own list of pests and diseases that represent risks to individual crops if introduced into the country. For Indonesia these are listed in the Lampiran Peraturan Menteri Pertanian Indonesia No, 31/Permentan/KR.010/7/2018 (website given in References). There are various animal pests from different taxa listed by the Indonesia Ministry of Agriculture through the Agriculture Ministerial Regulation body. These pests include insects, mites, snails and slugs, and nematodes, however, although there are as yet no snail or slug pests known for oil palm. Diseases include bacteria, fungi, viruses and other microorganisms. The official list shows how these pests and diseases may be carried, e.g. through seeds, and other plant parts. These lists are up-dated regularly and the most recent ones need to be consulted when considering importation of oil palm materials. There are stringent regulations on the movement of oil palm materials between countries and particularly between Africa, South America and other continents.

12.2   Pest and Disease Control

The regulation that governs the plant (seed and seedling) quarantine outside the entry and exit location is the Agriculture Ministerial Regulation No. 38/Permentan/OT.140/3/2014 (Peraturan Menteri Pertanian Republik Indonesia No. 38/Permentan/OT.140/3/2014). The law consists of some location requirements that relate to: (i) visual inspection; (ii) laboratory inspection; (iii) biological agent inspection; (iv) location of isolation and observation; (v) location for treatment; (vi) location of control action; and (vii) location of discard action. This manual only examines the requirements that relate to the nursery (requirements iv, v, vi and vii).

12.3   Location: Isolation and Observation (Singmat Area)

The location of an isolation and observation area can be a closed or an open facility.

Closed facility

A closed facility is required when the location is surrounded by the same vegetation (oil palm) and is only intended for the pre-nursery (Figs 12.1, 12.2). The closed facility may be a glasshouse, screenhouse or polyhouse. The requirements for a glass/screen/polyhouse consist of the following:

- **Doors.** Double doors with aluminium and wire netting at the second door. There should be a changing room, complete with boots/footwear exchange. Disinfection takes place between the first and second door.
- **Floor.** A concrete floor is suggested. If the surface is soil, this must be coated with a waterproof material. The drainage holes should be

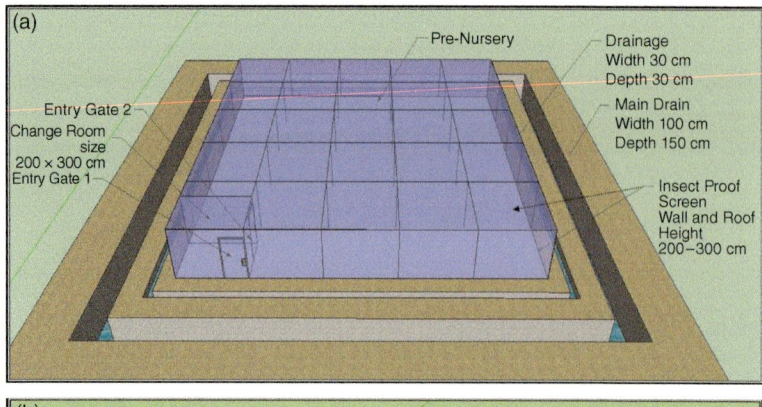

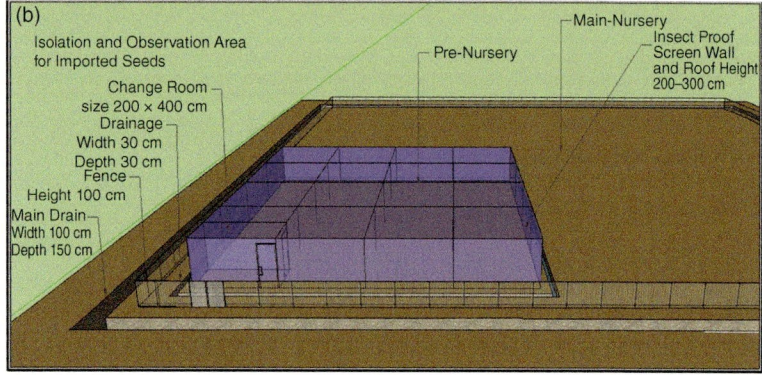

Fig. 12.1. Isolation and observation area layout for oil palm nursery quarantine regulation for oil palm around: (a) pre-nursery (3 months); (b) pre-nursery and main-nursery in one area.

Fig. 12.2. Closed facility for oil palm pre-nursery.

covered by wire netting to prevent the entry of animals, and an isolated trench with water must surround the glass/screen/polyhouse to restrict ingress of insects and animals.

- **Structure design.** The design may vary depending on the plant material and the location where the house is built. The roof may be curved or angled and the building frame should be made of galvanised or other stainless metal.
- **Roof.** The roof is best made of clear fibreglass or polythene with an insect-proof screen. In order to manage sunlight/shading, paranet should also be installed.
- **Wall.** An insect-proof screen for the wall is suggested. If the wall is polythene or polycarbonate, this should be coated with an insect-proof screen.
- **Closure of perforated structures.** Every perforated structure (e.g. exhaust fan) should be covered with an insect-proof screen (40–60 mesh). The screen must be made of stainless steel/phospher bronze.
- **Lighting/heating/cooling.** The greenhouse should have sufficient lighting for plant growth. In cold environments, the installation should be equipped with heating distributed to all areas. In hot climates the installation should be equipped with a cooling system (e.g. air conditioner).
- **Watering.** The greenhouse should have a suitable watering installation.
- **Cleaning and pot washing.** The greenhouse should be equipped with facilities for cleaning, washing and disinfecting pots separately from the main installation.
- **Potting area.** The greenhouse should be equipped with a potting area (complete with growth media and storage). This must be sited separately from the main installation.
- **Soil treatment area.** The greenhouse should be equipped with facilities for soil and other growth media sterilisation.
- **Incinerator.** The greenhouse should be equipped with an incinerator, located separately from the main installation.
- **Benches.** The greenhouse should be equipped with benches on which to place plant racks. The minimum distance between benches is 0.75–1.00 m. The racks are best made of stainless steel.
- **Waste management.** The greenhouse should be equipped with a water waste management facility for solid and liquid waste. Solid waste can be discarded after being sterilised, while liquid waste should be collected and drained into a septic tank.
- **Standard operating procedures (SOPs).** The greenhouse should have SOPs in place that cover all activities in the isolation and observation location.
- **Operator.** The greenhouse should be operated by a trained operator.

Open facility

An open facility could be an isolated area of land/plantation or an isolated nursery. An open facility is required when the location is free from the same vegetation (oil palm) (Figs 12.3; 12.4, 12.5).

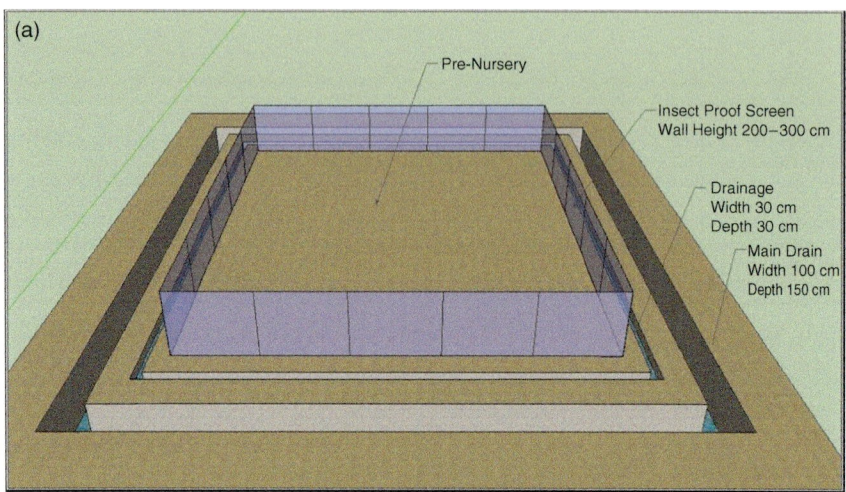

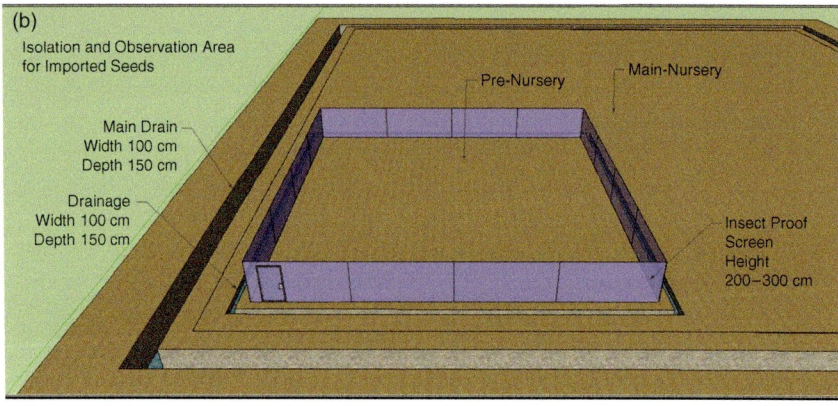

Fig. 12.3. Isolation and observation area layout for oil palm quarantine nursery, free of oil palm vegetation: (a) pre-nursery (3 months); (b) pre-nursery and main-nursery in one area.

This location should fulfil a number of requirements.

- **Location.** The location should accommodate and be suitable for seedling and ramet growth and development. The location should be isolated and have clear boundaries in terms of ownership and administration. There should be no plants of a similar species in the location.

Fig. 12.4. Open facility for oil palm pre-nursery.

Fig. 12.5. Open facility for oil palm main nursery.

- **Security.** The location must have a perimeter fence. Warning signs must be installed at the nursery entrance, and the nursery should have a security person in charge.
- **Soil/growth media treatment.** Soil/growth media tests and treatments should be conducted before the area is used.
- **Barrier.** A natural barrier in the form of wild plants that are not the plant pest organism's (PPO) host plants should be sufficient. An artificial barrier can be made of polythene plastic with a height of 3 m. *Daincha* or *Sesbania* may also be used as a barrier.
- **Watering.** The nursery should have a sufficient watering installation. It is better not to use overhead (sprinkler) watering systems to prevent OPT spread through water splash.
- **Drainage.** The nursery should have a good drainage system to prevent flooding. It should also have a good supply of water, e.g. a reservoir.
- **Sanitation.** The nursery should be free from weeds and wild plants. Any weed and wild plant remains should be discarded before the nursery is built.
- **Incinerator.** The nursery should be equipped with an incinerator.
- **SOPs.** The nursery is to have SOPs that cover all activities in the isolation and observation location.
- **Nursery coordinator.** The owner should designate a person who will coordinate with quarantine officials.
- **Nursery coordinator assistant.** The owner should designate a person to assist the process.

12.4   Location for Intervention Treatments

In order to mitigate the spread of OPT, some treatments are needed, such as fumigation and heat treatment. The location requirements are outlined below.

Fumigation location

- The location must be extensive and far from any settlements. The location should also fulfil the fumigation technical standard.
- The location is easy to reach and has access to transportation.
- The location has electricity and watering facilities.
- The location must be protected from strong winds and rain.
- The location must have sufficient ventilation and lighting.
- The location must have safe environmental conditions.
- The location must be free from flooding or water puddling.
- The location must have a gas-tight, flat and clean floor.

- The location must have a facility that will guarantee the impossibility of OPT reinvestment.
- For the phosphine application, the location must have a phosphine deactivated area.

12.5   Heat Treatments

Heat treatments are divided into two types based on the heat source media.

Air heat treatment

The location must be specifically designed to hold heat in order to reach a certain temperature. The requirements of an air heat treatment facility are as follows:

- The facility has strong walls and is able to withstand heat and not be easily corroded.
- The facility has a strong ceiling that is made of metal and is able to withstand heat. The ceiling must also be fireproof.
- The facility must have a concrete floor and be built higher than ground level.
- The facility must have a strong door made of metal and coated with an insulator. The door must also withstand heat and be easily opened and closed.
- The main room structure must be clean with no cracks.
- The heater must be able to produce heat at a certain temperature and time. The heat produced should be spread evenly and controlled automatically.
- The facility must be equipped with a sensor that has readability capacity with minimum 0.1 of scale. The sensor must also be connected to the monitor panel and be able to read the measurement results that consist of:
  o   dry bulb and wet bulb thermometers;
  o   three thermocouples complete with heat resistant wire; and
  o   probe thermocouple.

The installation must also be equipped with a blower, damper, heating valve, sprayer and panel control.

Vapour heat treatment

- The location, construction and facility design must pay attention to safety norms.
- The facility should be integrated with the packing area.

- The facility must be located in a safe area complete with security that prevents OPT reinvestment.
- The facility must be equipped with netting on all ventilation to prevent OPT entry.
- All the measurement equipment must be calibrated routinely.
- The boiler capacity must be sufficient to increase the room temperature to 50–52°C. The boiler should be able to produce a temperature between 46–48°C and to maintain that temperature for up to four hours.
- The facility should have a portable or permanent temperature sensor to monitor the temperature during application.
- All temperature changes must be recorded in one system.
- The facility should have a waste area to place commodity damaged during application.

12.6   Location for Control Actions

The approved location for control actions could be an open area, a separate building and/or part of the building such as a greenhouse/screenhouse. The facility should be able to accommodate all carrier media subject to control actions. The requirements for the location of discard actions are as follows:

- must be far from settlements and offices;
- must have a permanent incinerator in an open area of 400 m²;
- must have a permanent fence with ± 2 m of height;
- must be free from wild vegetation;
- must be equipped with a fire extinguisher;
- must have electricity and water installation;
- must have work safety and health facilities.

12.7   Inspection

Inspection of the quarantine nursery is governed by the regulation of the Decision of Head of Indonesia Agricultural Quarantine Agency no. 605/2012 (Keputusan Kepala Badan Karantina Pertanian no. 605/Kpts/HK.310/L/05/2012). In this regulation, the quarantine plants must be observed for six months before being released or detained. Inspections are conducted every month by the Quarantine Agency officers.

References

Crop Protection Compendium. https://www.cabi.org/cpc

Flood, J., Hasan, Y., Turner, P.D. and O'Grady, E.B. (2000) The spread of *Ganoderma* from infective sources in the field and its implications for management of the disease in oil palm. In: Flood, J., Bridge, P.D. and Holderness, M. (eds) Ganoderma *Disease of Perennial Crops*. CAB International, Wallingford, UK, pp. 101–112.

Hogenhout, S.A., Oshima, K, Ammar, E., Kakizawa, S., Kingdom, H.N. and Namba, S. (2008) Phytoplasmas: bacteria that manipulate plants and insects. *Molecular Plant Pathology* 9(4), 1–21.

Hushiarian, R., Yusof, N.A. and Dutse, S.W. (2013) Detection and control of *Ganoderma* boninense: strategies and perspectives. *Springer Plus* 2(555), 1–12.

Kushairi, D. (2012) Welcome remarks. In: Kien, W.C. (ed.) *Proceedings of 2012 the International Society for Oil Palm Breeders (ISOPB) International Seminar on Breeding for Oil Palm Disease Resistance, 21–24 November 2012*. Bogota, Colombia.

Rahmaningsih, M., Virdiana, I., Bahri, S., Anwar, Y., Forster, B.P. and Breton, F. (2018) *Nursery Screening for* Ganoderma *Response in Oil Palm Seedlings: A Manual*. In: *Techniques in Plantation Science*. Forster, B.P. and Caligari P.D.S. (eds). CAB International, Wallingford, UK, p. 50.

Rajanaidu, N., Kushairi, A., Din, A., Noh, A., Norziha, A. and Ainul, M. (2012) Breeding materials for disease resistance in oil palm – current status. In: Kien, W.C. (ed.) *Proceedings of 2012 the International Society for Oil Palm Breeders (ISOPB) International Seminar on Breeding for Oil Palm Disease Resistance, 21–24 November 2012*. Bogota, Colombia, Paper 1, pp. 1–2.

Rakib, M.R.M., Bong, C.F.J., Khairulmazmi, A. and Idris, A.S. (2014) Genetic and morphological diversity of *Ganoderma* species isolated from infected oil palms (*Elaeis guineensis*). *International Journal of Agriculture and Biology* 16(4), 691–699.

Turner, P.D. (1981) *Oil Palm Disease and Disorders*. Oxford University Press, Oxford.

Index

Page numbers in **bold** type refer to figures and tables.